ADVANCED LEVEL

CHEMISTRY
RESOURCE PACK

For AQA

ROY FARROW • DEREK GIBBENS
MARTIN STIRRUP • RAY VOWLES

Heinemann

Heinemann Educational Publishers,
Halley Court, Jordan Hill, Oxford OX2 8EJ
Part of Harcourt Education Ltd.

Heinemann is the registered trademark of Harcourt Education Ltd.

© R Farrow, D Gibbens, M Stirrup and R Vowles 2000

© Heinemann Educational Publishers 2000. The material in this publication is copyright. The duplicating masters may be photocopied for one-time use as instructional material in a classroom by a teacher, but they may not be copied in unlimited quantities, kept on behalf of others, passed on or sold to third parties, or stored for future use in a retrieval system. If you wish to use the material in any way other than that specified you must apply in writing to the publisher.

First published 2000

ISBN 0 435 58135X

05 04 03
10 9 8 7 6 5 4 3

Development Editor Paddy Gannon
Edited by Helen Roberts

Designed and typeset by AMR Ltd

Illustrated by Art Construction and Chartwell Illustrators

Original illustrations © Heinemann Educational Publishers 2000

Cover design by NB Designs

Printed and bound in the UK by Athenaeum Press Ltd.

Tel: 01865 888058 www.heinemann.co.uk

Contents

		Page No.
How to use this pack		iv
Key skills matching chart		1
Worksheets		
1	Atom size	3
2	Isotopes	4
3	Mass spectrometer	5
4	Working out masses	6
5	Electron energy levels	7
6	Filling the levels	8
7	Electronic structure	9
8	Evidence for energy shells	10
9	Trends down a Group	11
10	Ionic and covalent bonding	12
11	States of matter (1)	13
12	States of matter (2)	14
13	Shapes of molecules and ions	15
14	Group II elements	16
15	Period 3 (1)	17
16	Compounds of Group II elements (1)	18
17	Compounds of Group II elements (2)	19
18	Period 3 (2)	20
19	Relative atomic and molecular mass	21
20	Avogadro's constant and moles (1)	22
21	Avogadro's constant and moles (2)	23
22	Empirical and molecular formulae	24
23	Balancing equations	25
24	Reacting masses	26
25	Reacting masses and volumes	27
26	Reacting volumes	28
27	The ideal gas equation (1)	29
28	The ideal gas equation (2)	30
29	Molarity and solutions	31
30	Volumetric analysis (1)	32
31	Volumetric analysis (2)	33
32	Enthalpy change and calorimetry	34
33	Calorimetry and using Hess's Law	35
34	Calculations using Hess's Law	36
35	Thermochemical calculations	37
36	Collision theory and rate of reaction (1)	38
37	Collision theory and rate of reaction (2)	39
38	Factors influencing reaction rate (1)	40
39	Factors influencing reaction rate (2)	41
40	Maxwell–Boltzmann distribution	42
41	Catalysts	43
42	Establishing equilibria	44
43	Applying Le Chatelier's Principle	45

		Page No.
44	A gaseous equilibrium	46
45	Production of ethanol	47
46	Production of ammonia	48
47	Production of sulphuric acid	49
48	Oxidation and reduction	50
49	Oxidation states (1)	51
50	Oxidation states (2)	52
51	What's in a name?	53
52	Oxidation states and Redox	54
53	Simple half-equations	55
54	Constructing whole equations	56
55	Trends	57
56	Halogens are oxidising agents	58
57	Halides are reducing agents	59
58	Which halide?	60
59	Halogens in water	61
60	Value for money	62
61	Principles of extraction	63
62	Aluminium and titanium	64
63	Methods of extraction	65
64	Preparing reports	66
65	Formulae of organic compounds	67
66	Structural isomerism	68
67	Empirical, molecular and structural formulae	69
68	Naming more organic compounds	70
69	Fractional distillation and cracking	71
70	Industrial applications	72
71	Combustion of fuels	73
72	Petroleum	74
73	Unsaturated hydrocarbons	75
74	Breaking the double bond	76
75	Major and minor products	77
76	Hydrogenation	78
77	Alcohols	79
78	Haloalkanes (1)	80
79	Haloalkanes (2)	81
80	Haloalkanes (3)	82
81	Alcohols and dehydration	83
82	Elimination and other organic reactions	84
83	Ethanol production and reactions	85
84	Reactions of epoxyalkanes	86
85	Classification and reactions	87
Answers to worksheet questions		88
Answers to student book 'test yourself'		107
Periodic Table		123

How to use this pack

This resource pack accompanies the AS Chemistry book for AQA. It provides photocopiable material covering the three modules: Atomic structure, bonding and periodicity; Foundation physical and inorganic chemistry; and Foundation organic chemistry.

The pack contains 85 worksheets, which have been designed to provide opportunities to practise key skills and to meet assessment objectives. A number of approaches have been used to develop and test knowledge and understanding, application, analysis and evaluation, as well as experiment and investigation.

Some worksheets have been designed to extend students beyond the specification, and some cover content from two or more of the AS modules.

A key skills matching chart maps the opportunities for practising application of number, communication and IT.

At the end of this pack there are answer sections which cover both the worksheets and the 'test yourself' questions from the student book.

There is a student book for the A2 year of the specification which also has a resource pack. These cover the two compulsory A2 modules: Further physical and organic chemistry, and Thermodynamics and further inorganic chemistry.

Key skills

Unit	Worksheet		Communication				Information technology			Application of number	
		3.1a	3.1b	3.2	3.3	3.1	3.2	3.3	3.1	3.2	3.3
Module 1 – Atomic structure, bonding and periodicity											
1 Atomic structure	1 Atom size			●						●	
	2 Isotopes			●						●	
	3 Mass spectrometer			●					●		
	4 Working out masses			●			●		●		
	5 Electron energy levels			●					●	●	
	6 Filling the levels			●							
	7 Electronic structure			●							
	8 Evidence for energy shells			●			●				
	9 Trends down a Group			●			●				
2 Bonding	10 Ionic and covalent bonding			●							
	11 States of matter (1)			●							
	12 States of matter (2)			●							
	13 Shapes of molecules and ions			●							
3 Periodicity	14 Group II elements			●			●	●	●		
	15 Period 3 (1)			●							
	16 Compounds and Group II elements (1)			●						●	
	17 Compounds and Group II elements (2)			●				●		●	●
	18 Period 3 (2)			●	●		●		●		●
4 Amount of substance	19 Relative atomic and molecular mass								●	●	
	20 Avogadro's constant and moles (1)								●	●	
	21 Avogadro's constant and moles (2)									●	
	22 Empirical and molecular formulae									●	
	23 Balancing equations										
	24 Reacting masses									●	
	25 Reacting masses and volumes									●	
	26 Reacting volumes									●	
	27 The ideal gas equation (1)								●	●	
	28 The ideal gas equation (2)								●	●	
	29 Molarity and solutions									●	
	30 Volumetric analysis (1)									●	
	31 Volumetric analysis (2)									●	
Module 2 – Foundation physical and inorganic chemistry											
5 Energetics	32 Enthalpy change and calorimetry			●						●	
	33 Calorimetry and using Hess's Law			●						●	
	34 Calculations using Hess's Law			●						●	
	35 Thermochemical calculations			●						●	
6 Kinetics	36 Collision theory and rate of reaction (1)				●						
	37 Collision theory and rate of reaction (2)									●	●
	38 Factors influencing reaction rate (1)									●	●
	39 Factors influencing reaction rate (2)									●	●
	40 Maxwell–Boltzmann distribution										
	41 Catalysts		●	●	●	●	●	●			
7 Equilibria	42 Establishing equilibria			●						●	
	43 Applying Le Chatelier's Principle			●							●

Unit	Worksheet	Communication				Information technology			Application of number		
		3.1a	3.1b	3.2	3.3	3.1	3.2	3.3	3.1	3.2	3.3
	44 A gaseous equilibrium			●						●	
	45 Production of ethanol			●							
	46 Production of ammonia			●							
	47 Production of sulphuric acid			●							
8 Redox reactions	48 Oxidation and reduction			●							
	49 Oxidation states (1)			●							
	50 Oxidation states (2)			●						●	
	51 What's in a name?			●							
	52 Oxidation states and Redox			●						●	
	53 Simple half-equations			●							
	54 Constructing whole equations			●							
9 Group VIII, the halogens	55 Trends			●							
	56 Halogens are oxidising agents			●							
	57 Halides are reducing agents			●							
	58 Which halide?			●							
	59 Halogens in water			●							
	60 Value for money			●			●			●	
10 Extraction of metals	61 Principles of extraction			●							
	62 Aluminium and titanium			●							
	63 Methods of extraction			●							
	64 Preparing reports	●	●	●	●						
Module 3 – Foundation organic chemistry											
11 Nomenclature and isomerism	65 Formulae of organic compounds									●	
	66 Structural isomerism								●	●	●
	67 Empirical, molecular and structural formulae										
	68 Naming more organic compounds										
12 Petroleum and alkanes	69 Fractional distillation and cracking			●						●	
	70 Industrial applications			●						●	
	71 Combustion of fuels			●							
	72 Petroleum			●							
13 Alkenes	73 Unsaturated hydrocarbons			●							
	74 Breaking the double bond			●							
	75 Major and minor products			●							
	76 Hydrogenation			●							
	77 Alcohols			●							
14 Haloalkanes	78 Haloalkanes (1)			●							
	79 Haloalkanes (2)			●							
	80 Haloalkanes (3)			●							
15 Alcohols	81 Alcohols and dehydration			●						●	
	82 Elimination and other organic reactions			●						●	
	83 Ethanol production and reactions			●						●	
	84 Reactions of epoxyalkanes			●						●	
	85 Classification and reactions			●						●	

Atom size

How many atoms are there on the head of a pin?

① Iron atoms have a diameter of approximately 2.5×10^{-10} m. A pin head has a diameter of approximately 1 mm (1×10^{-3} m). So the number of atoms across the diameter of the pin head would be ...?

② And the approximate number of atoms covering the surface of the pin head would be ...? (Simply square the linear number for an approximate answer.)

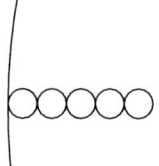

How many protons?

③ The mass of a proton is 1.672×10^{-27} kg. A hydrogen atom has one proton and no neutrons. How many protons are there in 1 g of hydrogen? (Find the answer to 1 s.f. – ignore the mass of the electrons.)

④ How many hydrogen atoms are there in 1 g of hydrogen?

⑤ a What is the charge on the nucleus of an atom of:
   i carbon ($Z = 6$)
   ii neon ($Z = 10$)
   iii aluminium ($Z = 13$)
   iv titanium ($Z = 22$)?
  b What is the total charge on an atom of each of these elements?

⑥ Copy and complete this table:

Element	Atomic number	Mass number	Number of protons	Number of neutrons	Number of electrons
Sulphur		32	16		
Calcium	20		20		
Iron		56			26
Lead	82	207			
Uranium			92	146	

Atomic number (Z) and mass number (A)

⑦ a What do the Z and A values shown tell you about the three elements Xa, Ya and Za? Give as much detail as you can.

    A value   40   40   40
    Element   Xa   Ya   Za
    Z value   18   19   20

  b Will their chemical properties be the same or different?

© Farrow R, Gibbens D, Stirrup M and Vowles R, 2000. AS Chemistry for AQA – Resource Pack, Heinemann

Isotopes

1 a What do the Z and A values shown tell you about the atoms of Xb, Yb and Zb? Give as much detail as you can.

A value	20	21	22
Element	Xb	Yb	Zb
Z value	10	10	10

b How would their chemical properties compare with one another?

2 a What is the difference between $^{16}_{8}O$ and $^{18}_{8}O$?

b O-18 is rare (0.2% of air). If you could get oxygen gas containing only O-18 atoms, how much denser than ordinary air would it be?

c What is the difference between $^{1}_{1}H$ and $^{2}_{1}H$ (often written as $^{2}_{1}D$)?

d H-2 (D-2) is rare (0.015% of hydrogen). If you could get hydrogen gas containing only D-2 atoms, how much denser than ordinary hydrogen would it be?

e Water made from H-2 atoms (D_2O) is used in nuclear reactors. How much denser than ordinary water would it be? Explain why this is usually referred to as 'heavy water'.

Isotope properties

3 a Radon gas diffuses into houses on Dartmoor and in Derbyshire from the surrounding granite. Which diffuses in faster, Rn-220 or Rn-224?

b H-2 (D) is separated from H-1 by concentration in hydrogen sulphide – stink bomb gas. If a stink bomb containing H_2S and D_2S was dropped in your vicinity, which molecules would you smell first? Explain your answer.

c U-235 is separated from U-238 by combining it with fluorine to make uranium hexafluoride (UF_6), boiling this compound and then diffusing the gas through a porous membrane. Molecules containing which uranium isotope will diffuse faster?

Mass spectrometer

①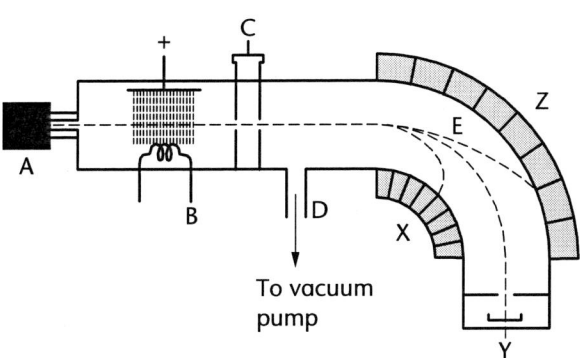

Sketch this simple diagram of a mass spectrometer and label:
- the vaporisation chamber
- the electron gun
- the ion accelerator
- the deflection zone
- the detector.

Briefly annotate your diagram to explain the function of each part.

② The spectrometer is set up so that a certain ion with a single positive charge is detected at Y. Would similar ions still be detected at Y or be deflected towards X or Z if:

a the strength of the magnetic field is increased
b the acceleration is increased
c the ion is given a double positive charge
d the temperature in the vaporiser is increased
e the strength of the magnetic field is decreased
f a single-charged ion of an isotope with a higher mass number passes through?

③ Look at the mass spectrum for potassium (single charge ions). It is set so that the y-axis gives the percentage of each isotope.

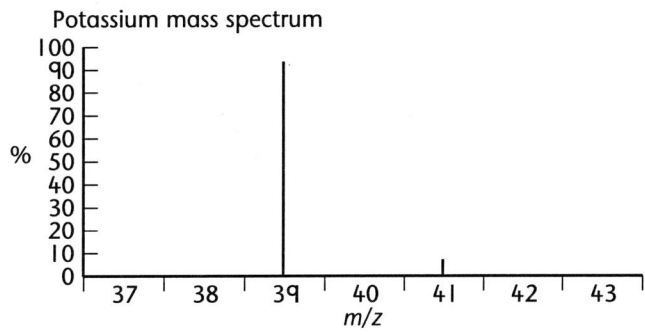

a What is the mass/charge ratio of the commonest ion?
b What is the mass number of this ion?
c What percentage of potassium atoms have this mass number?
d What is the mass number of the less common isotope of potassium?
e What percentage of potassium atoms have this mass number?

Module 1 — Atomic structure, bonding and periodicity — 1 Atomic structure

Working out masses

4

① Copper has two isotopes: $^{63}_{29}$Cu (69.1%) and $^{65}_{29}$Cu (30.9%).
Calculate the relative atomic mass of copper from these figures, showing your working.

② **a** The table below shows a section from a spreadsheet designed to work out relative atomic mass numbers from up to three isotope abundance percentages. Copy this onto a suitable spreadsheet and use your answer to Question 1 to help you work out a suitable formula to go into cell H3, which will give the answer as shown. Remember that an extra term will be needed to allow for the third possible isotope.

A	B	C	D	E	F	G	H
1	isotope 1		isotope 2		isotope 3		overall A_r
2 element	A	%	A	%	A	%	
3 Ga	69	60.4	71	39.6	0	0	69.79
4 Ag	107	51.35	109	48.65	0	0	
5 Si	28	92.18	29	4.71	30	3.12	
6 Ne	20	90.92	21	0.06	22	8.82	
7 Mg	24	78.6	25	10.11	26	11.29	

b Now replicate this formula for cells H4 to H7, to find the overall relative atomic masses of Ag, Si, Ne and Mg.

③

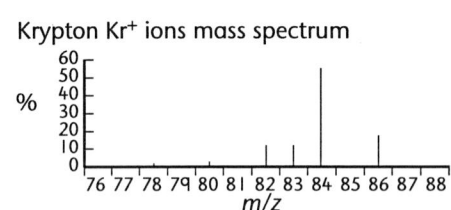
Krypton Kr⁺ ions mass spectrum

a Look at the mass spectrum of krypton (Kr⁺) ions shown above. List the isotopes shown with their A numbers in order of descending abundance.
b Which integer value is the A number for krypton likely to be closest to?
c A small peak was also seen at $m/z = 42$. What might this represent?

④ Bromine has two isotopes of A 79 and A 81, which occur naturally in roughly equal proportions. Bromine forms Br$_2$ molecules.

a Explain each of the peaks on the mass spectrum shown below.

Bromine mass spectrum

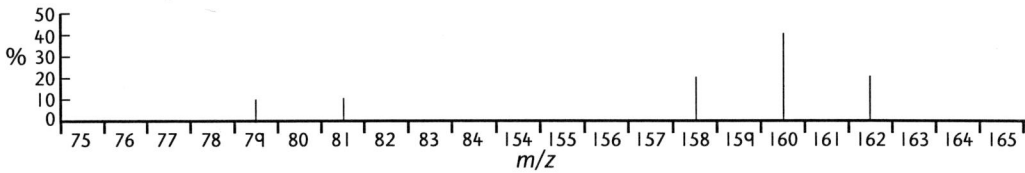

b Why is the peak at $m/z = 160$ twice as high as those at 158 and 162?

© Farrow R, Gibbens D, Stirrup M and Vowles R, 2000. AS Chemistry for AQA – Resource Pack, Heinemann

Module 1 — Atomic structure, bonding and periodicity — 1 Atomic structure

Electron energy levels

5

① The atomic number (Z) of calcium is 20. Draw simple electronic structure diagrams for a calcium atom and a calcium (Ca^{2+}) ion based on your knowledge from GCSE.

② Draw simple electron configuration diagrams for:
  a $_{17}$Cl atom
  b $_{17}$Cl$^-$ ion
  c $_{13}$Al^{3+} ion
  d $_7$N^{3-} ion
  e $_{14}$Si atom.

③ At GCSE, the structure of calcium is written as Ca 2, 8, 8, 2. Explain what you now understand by 2, 8, 8, 2 in terms of principal energy levels and sub-levels. How may the 8s be subdivided?

④ Copy and complete this table showing the number of electrons that can exist in each sub-level of each principal energy level. Add the maximum number of electrons for each principal level. Mark X the sub-levels that do not exist.

Principal energy level	Sub-level s	Sub-level p	Sub-level d	Sub-level f	Total in principal energy level
1		X	X		2
2		6			
3					
4					

Electron levels and the Periodic Table

⑤ The layout of the Periodic Table follows the pattern of electron arrangements in the various energy levels and sub-levels.

  a Sketch this outline of the Periodic Table and label the blocks that correspond to the filling of sub-levels s, p and d.

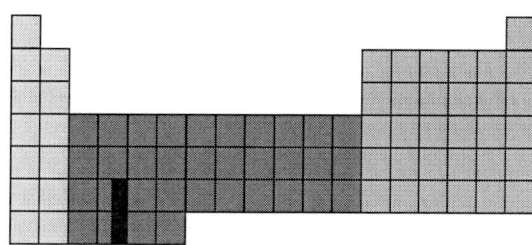

  b Explain the width of these s, p and d blocks.
  c At ▮ (levels 6 and 7) there are missing rows of elements that correspond to the filling of the sub-level f. How many elements are missing from each row?

⑥ Which group of the Periodic Table is characterised by:
  a having its highest p sub-level full?
  b having only 1 electron in its highest s sub-level?
  c having 5 electrons in its highest p sub-level?

© Farrow R, Gibbens D, Stirrup M and Vowles R, 2000. AS Chemistry for AQA – Resource Pack, Heinemann

Module 1 — Atomic structure, bonding and periodicity — **1 Atomic structure**

Filling the levels

(1) At GCSE, the structure of calcium is written as Ca 2, 8, 8, 2. This is now written to show how the s and p sub-levels are filled:

Ca $1s^2 2s^2 2p^6 3s^2 3p^6 4s^2$ (Notice here that it is the superscript numbers that add up to 20.)

In the notation $3p^6$ explain what the 3, p and 6 tell us.

(2) Write out the electronic structure of: $_8O$, $_{10}Ne$, $_{11}Na$, $_{17}Cl$ and $_{19}K$ in this way.

(3) You might expect these to fill in order: 1s → 2s → 2p → 3s → 3p → 3d → 4s → 4p and so on, but the energy levels overlap slightly with 4s lower than 3d, so 4s fills before 3d giving the sequence 3p → 4s → 3d.
Use this idea to work out for yourself the full electronic structure of: $_{22}Ti$, $_{25}Mn$ and $_{27}Co$.

Electronic structure shorthand

(4) The electron structure of argon is $1s^2 2s^2 2p^6 3s^2 3p^6$, while calcium is $1s^2 2s^2 2p^6 3s^2 3p^6 4s^2$.

A shorthand form for calcium is therefore $[Ar]4s^2$.

Use this shorthand method to write electronic structures for: K, V, Ni, As and Br.

(5) What element is shown by each of these structures?

a $[Ne]\, 3s^2 3p^4$
b $[Ne]3s^2$
c $[Ar]\, 3d^6 4s^2$
d $[Kr]\, 5s^2$
e $[Kr]\, 4d^2 5s^2$
f $[Xe]\, 6s^2$

(6) a What is **anomalous** about the structure of Cr $1s^2 2s^2 2p^6 3s^2 3p^6 3d^5 4s^1$ and Cu $1s^2 2s^2 2p^6 3s^2 3p^6 3d^{10} 4s^1$?

b Give a simple explanation for this anomalous shell filling.

(7) Silver has 18 more protons than copper. Suggest its likely electron configuration.

Electronic structure

1 Each energy sub-level may be split into orbitals, each of which can hold up to two electrons.

Copy and complete this table showing how many electrons and orbitals may be found in each sub-level.

Energy sub-level	Maximum number of electrons	Number of orbitals
s		
p		
d		
f		

2 Copy and complete this table showing the total number of electrons and orbitals in each principal energy level.

Principal energy level	Maximum number of electrons	Number of orbitals
1		
2		
3		
4		

3 What is the simple mathematical relationship between the number of the principal energy level and the number of orbitals that it contains?

4 Use Hund's rule to show how electrons fill the orbitals for:

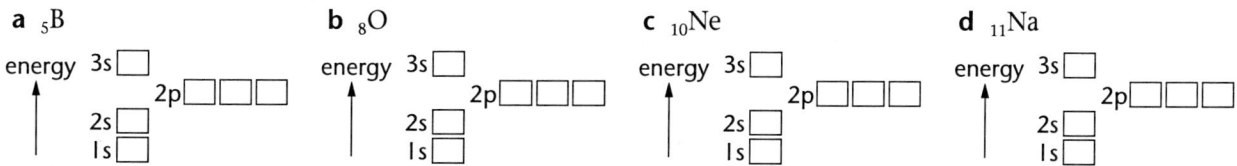

a $_5$B   b $_8$O   c $_{10}$Ne   d $_{11}$Na

Electronic structure of ions

5 Show how the orbitals are filled for:
a $_9$F$^-$ ion   b $_{11}$Na$^+$ ion   c $_{12}$Mg^{2+} ion

6 $_{26}$Fe^{3+} 1s^{2}2s^{2}2p^{6}3s^{2}3p^{6}3d^5 is more stable than $_{26}$Fe^{2+} 1s^{2}2s^{2}2p^{6}3s^{2}3p^{6}3d^6 because it has a half-full d-shell.

  a Write out full electron configurations like this for a $_{25}$Mn atom, a $_{25}$Mn^{2+} ion and a $_{25}$Mn^{3+} ion.

  b Explain, with reasons, which ion is likely to be more stable.

7 a $_{29}$Cu has two common ions, Cu$^+$ and Cu^{2+}. From your experience of common compounds such as copper sulphate, which of these ions appears to be more stable?

  b Write out full electron configurations for a $_{29}$Cu atom, a $_{29}$Cu$^+$ ion and a $_{29}$Cu^{2+} ion. Explain why copper ions are often said to show anomalous stability.

© Farrow R, Gibbens D, Stirrup M and Vowles R, 2000. AS Chemistry for AQA – Resource Pack, Heinemann

Evidence for energy shells

1 Our understanding of electronic structure comes from the energy needed to remove successive electrons from an atom. Look at this table of energy values for calcium. These show the energy needed to remove successive electrons. Note that the table is starting from the outside of the atom and working in. (Don't worry about the units used for the moment, as we are only interested in the pattern.)

No of e	IE/kJ mol⁻¹	log₁₀
1	590	2.77
2	1145	3.06
3	4912	3.69
4	6474	3.81
5	8145	3.91
6	10496	4.02
7	12320	4.09
8	14207	4.15
9	18192	4.26
10	20385	4.31
11	57048	4.76
12	63333	4.80
13	70052	4.85
14	78792	4.90
15	86367	4.94
16	94000	4.97
17	104900	5.02
18	111600	5.05
19	494790	5.69
20	527759	5.72

   a Plot a graph of the number of electrons removed against the log of the energy needed. (Using the log values displays the pattern more clearly.) You may wish to use a spreadsheet to do this.

   b How does the pattern support the idea of four principal electron energy levels ('shells')?

   c Now plot a graph of the actual energy level against number for electrons 3–10. Can you spot evidence for the existence of sublevels within this principal energy level?

2 The table below shows the first ionisation energies for elements Li–Ne and Na–Ar.

   a Plot a line graph showing a separate line for each Period. (You may do this by hand or using a spreadsheet.)

Period 2	Li	Be	B	C	N	O	F	Ne
Energy/kJ mol⁻¹	519	900	799	1090	1400	1310	1680	2080
Period 3	Na	Mg	Al	Si	P	S	Cl	Ar
Energy/kJ mol⁻¹	494	736	577	786	1060	1000	1260	1520

   b Annotate your graph to highlight and explain:

      i the drop in value from Group 2 to 3 for each Period
      ii the drop in value from Group 5 to 6 for each Period
      iii the relative line position for Periods 2 and 3.

3 For each of the following element pairs, state which element you would expect to have the higher first ionisation energy, giving reasons for your answers.

   a rubidium ($_{37}$Rb) and strontium ($_{38}$Sr)
   b antimony ($_{51}$Sb) and bismuth ($_{83}$Bi)
   c calcium ($_{20}$Ca) and scandium ($_{21}$Sc)
   d arsenic ($_{33}$As) and selenium ($_{34}$Se)

© Farrow R, Gibbens D, Stirrup M and Vowles R, 2000. AS Chemistry for AQA – Resource Pack, Heinemann

Trends down a Group

① a Using the values given in the table below, plot graphs to show the trends of the first ionisation energies for elements going down Groups I, VI and VIII.

Group I		Group VI		Group VIII	
Element	Energy/kJ mol^{-1}	Element	Energy/kJ mol^{-1}	Element	Energy/kJ mol^{-1}
Li	519	O	1310	Ne	2080
Na	494	S	1000	Ar	1520
K	418	Se	941	Kr	
Rb		Te		Xe	1170
Cs	376	Po	812	Rn	1040

b Use these trendlines to estimate the missing values for Rb, Te and Kr.

c Explain why the values go down as you go down each Group.

d Explain why the values go up across each Period from Group VI to Group VIII.

② a Use the figures from the table opposite to plot the successive ionisation energies for aluminium. For clarity, use a spreadsheet and set the graph to display a logarithmic scale.

b Annotate the graph to clearly show the discontinuities shown in the line by the changes in energy sub-levels: from 3p to 3s, from 3s to 2p and from 2s to 1s.

c Which sub-level change is not shown clearly from this graph? Plot a more detailed graph from electrons 4 to 10 (plain scale) to locate this change.

Successive ionisation energies for aluminium/ kJ mol^{-1}	
1	578
2	1817
3	2745
4	11 578
5	14 831
6	18 378
7	23 296
8	27 460
9	31 862
10	38 458
11	42 655
12	201 276
13	222 313

③ a Use the figures given in the table below to plot a line graph of the final ionisation energies for the first 10 elements. (The energy needed to remove the final 1s electron.)

Final electron ionisation energy/kJ mol^{-1}	
H	1312
He	5251
Li	11 815
Be	21 007
B	32 828
C	47 278
N	64 362
O	84 080
F	106 437
Ne	131 435

b Explain the pattern shown by this graph.

c Predict the final ionisation energy for sodium.

© Farrow R, Gibbens D, Stirrup M and Vowles R, 2000. AS Chemistry for AQA – Resource Pack, Heinemann

Ionic and covalent bonding

1 Write brief notes, with a definition if appropriate, on each of the following:
(Refer to the student's book if you need some help.)

a ionic bonding
b covalent bonding
c co-ordinate bonding
d metallic bonding
e electronegativity
f polar bonding
g charge density
h polarising power of a cation
i permanent dipole-dipole forces
j induced dipole-dipole (van der Waals') forces
k hydrogen bonding
l ionic crystals
m molecular crystals
n giant covalent crystals (macromolecules)
o electron pair repulsion theory

Ionic and covalent compounds

2 By reference to the Periodic Table predict the formula of each of the following **ionic** compounds.

a caesium fluoride
b potassium oxide
c calcium oxide
d sodium nitride
e aluminium fluoride
f magnesium sulphide

3 Describe the bonding in each of the following. In each case give a full, detailed description with reasons.

a KCl
b Br_2
c IBr
d NH_4Cl
e HF
f $AlCl_3$
g H_3O^+
h MgF_2
i $BeCl_2$
j LiI

4 Predict which compound, in each of the following pairs has the higher melting point. Explain your reasoning.

a NaF and MgF_2
b Na_2O and MgO
c NaCl and $AlCl_3$

Consequences of bonding

5 Explain each of the following observations.
a Solid iron conducts electricity.
b Sodium has a higher melting point than potassium.
c Sodium has a lower melting point than magnesium.
d ICl has a higher melting point than Br_2.
e HF has a higher boiling point than HCl.
f Chlorine melts at –101 °C, but a temperature well above 1000 °C is needed to convert chlorine molecules into atoms.
g Solid sodium chloride does not conduct electricity, but when molten or dissolved in water, it does.
h Ammonia gas is very soluble in water.
i Methane has a very low boiling point (–162 °C).
j The electronegativity of neon is zero.
k One of the aluminium–chlorine bonds in $AlCl_4^-$ is formed differently from the other three.

© Farrow R, Gibbens D, Stirrup M and Vowles R, 2000. AS Chemistry for AQA – Resource Pack, Heinemann

States of matter (1)

1 a Ethanol (C_2H_5OH) boils at 79 °C whereas propane (C_3H_8), a compound with a similar molecular mass boils at –42 °C. Suggest a reason for this large difference.

b Another compound with a similar molecular mass is ethylamine ($C_2H_5NH_2$). How do you think its boiling point will compare with the boiling point of propane? Explain your answer.

2 What intermolecular forces must be overcome to:
- **a** melt ice
- **b** boil chloromethane (CH_3Cl)
- **c** vaporise iodine
- **d** boil liquid nitrogen?

3 Oil floats on water. From this comment, what conclusion can you draw concerning the ability of the molecules in oil to form hydrogen bonds?

4 Consider the information in the following table.

Substance	Melting point/°C	Boiling point/°C	Conduction of electricity: solid	liquid	in water
A	–78	–33	none	none	good
B	–117	79	none	none	none
C	1610	2230	none	none	none
D	850	1490	good	good	good
E	970	1550	none	good	good
F	–199	–189	none	none	none
G	114	184	none	none	none

Use the information in the table to identify:
- **a** a gas at room temperature which does not react with water
- **b** a gas at room temperature which reacts with water to give an ionic solution
- **c** an ionic solid at room temperature
- **d** a macromolecular solid
- **e** a metal which reacts with water
- **f** a solid at room temperature, with weak intermolecular forces of attraction
- **g** a liquid at room temperature.

The seven substances are carbon monoxide, silicon dioxide, ethanol, iodine, ammonia, barium chloride and calcium. Deduce which is which.

Module 1 — Atomic structure, bonding and periodicity — 2 Bonding

States of matter (2)

1 By considering the behaviour of the molecules in water, explain the difference between evaporation and boiling. Include in your answer the reason why a liquid can vaporise below its boiling point.

2 How would you expect the dissolving of sodium chloride in water to affect its:
   a freezing point
   b boiling point?

3 a When potassium chloride dissolves in water what forces must be overcome in:
   i the solid potassium chloride
   ii the water?
   b What forces enable the potassium chloride to dissolve in water?

4 Explain why much more energy is needed to convert water into water vapour than is needed to convert ice into water.

5 In the sodium chloride crystal structure each sodium ion is surrounded by six chloride ions, whereas in caesium chloride each caesium ion is surrounded by eight chloride ions. Account for this difference.

6 a Graphite electrodes are sometimes used in electrolysis. With reference to its structure, explain why graphite is a conductor of electricity.
   b Explain why diamond is a very hard substance but graphite is soft.
   c Both diamond and graphite have melting points above 3000 °C. Give a reason for this.

7 Consider the following statements concerning the behaviour of solids, liquids and gases.
   i Rapid randomly-moving particles with no attractive forces between them.
   ii Strong forces of attraction between particles which are vibrating about fixed points.
   iii Attractive forces between cations and electrons.
   iv It is a large molecule containing lots of covalent bonds.
   v Its structure contains delocalised electrons.
   vi Weak forces of attraction between the molecules.

   Choose the most suitable statement from the above list to explain each of the following statements.
   a Sodium is a metal at room temperature.
   b Diamond has a very high melting point.
   c Oxygen is a gas at room temperature.
   d Sodium chloride has a high melting point.
   e Graphite conducts electricity.
   f Iodine is a solid with a low melting point.

© Farrow R, Gibbens D, Stirrup M and Vowles R, 2000. AS Chemistry for AQA – Resource Pack, Heinemann

Shapes of molecules and ions

① Examine the following statements and say whether each one is true or false. Explain your reasoning fully, correcting any false statements.
  a $BeCl_2$ is a linear molecule.
  b BF_3 is a trigonal planar molecule.
  c The bond angle in PH_3 is 120°.
  d The bond angle in H_2O is 107°.
  e IF_5 is a trigonal bipyramidal molecule.
  f $[BF_4]^-$ is tetrahedral.
  g The shape of OF_2 is the same as H_2O.
  h $[SbF_6]^-$ is octahedral.
  i $[ICl_2]^+$ is linear.
  j $[ICl_4]^+$ is square planar.

② Deduce the shape of each of the following species. Suggest values for the bond angles.
  a $[NH_2]^-$ c $[ICl_2]^-$ e $[ICl_4]^-$ g $[PCl_4]^+$ i SCl_2
  b XeF_2 d $[TeF_5]^-$ f XeF_4 h BrF_3 j $[I_3]^-$

③ Hydrogen fluoride undergoes self-ionisation to give H_2F^+ and HF_2^- ions.
  a Deduce the shapes of these two ions.
  b Hydrogen fluoride reacts with boron trifluoride according to the equation:
     $2HF + BF_3 \rightarrow H_2F^+ + BF_4^-$
     i   What type of bond is formed to convert BF_3 into BF_4^-?
     ii What feature of BF_3 enables it to form this type of bond?
     iii How does the F—B—F bond angle change when this new bond is formed? Explain your answer.
     iv Suggest, with a reason, another molecule which could react with BF_3 in the same way.

④ In the vapour state aluminium chloride molecules dimerise to form Al_2Cl_6.
  a Describe the bonding between the molecules in the formation of the dimer.
  b Suggest how the Cl—Al—Cl bond angle changes as the dimer is formed.

Group II elements

The table below contains some of the data that you will need to use in order to answer the questions on this sheet. Two pieces of data have been left out so that you can work out their predicted values from Group trends.

Element	Atomic number	Atomic radius/nm	First ionisation energy/kJ mol^{-1}
Beryllium	4	0.112	900
Magnesium	12	0.160	740
Calcium	20		
Strontium	38	0.215	550
Barium	56	0.222	500

1. Plot a graph of atomic number (on the horizontal axis) against atomic radius.

    You may wish to do questions 1 and 4 on a PC with a spreadsheet.

2. What is the trend in atomic radius as you go down Group II?

3. Why does this happen?

4. On the same axes, also plot first ionisation energy against atomic number. You may have to use a different vertical scale.

5. Use the lines on your graph to predict the missing values for calcium.

    The real values are a little different from the predicted ones so just looking them up will not help you.

6. State and explain the trend in first ionisation energy as you go down Group II.

7. Say how you would **expect** the electronegativity values of the elements to vary as you go down the Group and explain why.

8. Explain how and why you would **expect** the melting points of the elements to change as you go down the Group.

9. **Why** is beryllium unlike the other members of Group II?

Period 3 (1)

Atomic radius

① This table contains the atomic radii of some of the elements in Period 3.

Na	Mg	Al	Si	S	Cl
0.156	0.136	0.125	0.117	0.104	0.099

Plot these values on a graph of atomic radius against atomic number and use your graph to predict a value for phosphorus.

② Explain the trend in atomic radius from left to right across the Period.

First ionisation energies

③ Sketch a graph to show how first ionisation energy varies across Period 3.

④ From what you know about periodic trends, sketch, on the same axes, a graph to show how you think first ionisation energy varies across Period 2. Label each sketch clearly.

⑤ Explain the difference between the first ionisation energies of aluminum and magnesium.

⑥ Explain the difference between the first ionisation energies of phosphorus and sulphur.

⑦ Which electron orbitals are being filled up as Period 3 is crossed?

⑧ If the third main energy level can hold up to 18 electrons, why are there only 8 elements in Period 3?

Electronegativities

⑨ Define the term electronegativity.

⑩ The box at the side shows the electronegativity values for the elements in Period 3. Plot a graph to show the variation in electronegativity across Period 3 and explain the trend.

Na	0.9
Mg	1.2
Al	1.5
Si	1.8
P	2.1
S	2.5
Cl	3.0

⑪ Why is there no value for argon?

Compounds of Group II elements (1)

Solubility of barium sulphate

Be	41.0
Mg	36.4
Ca	0.210
Sr	0.010
Ba	0.000266

① The solubilities of the Group II metal sulphates in g/100g water are shown in the box. Sketch a graph to show how these values change with atomic number.

② Work out the solubility of barium sulphate in $mol\,dm^{-3}$.

③ A precipitate of barium sulphate forms whenever the molar concentration of Ba^{2+} multiplied by the molar concentration of SO_4^{2-} exceeds $1.30 \times 10^{-10}\,mol^2\,dm^{-6}$.

If a solution of 0.03 M barium nitrate is added to an equal volume of sodium sulphate solution what is the minimum concentration of the sodium sulphate solution that will cause a precipitate to form?

> Hint – the concentration of each solution will be halved when equal volumes are mixed.

④ To what practical use can this be put in **qualitative** analysis?

⑤ A preparative laboratory needed a very pure form of sodium sulphate. A supplier claimed that their analytical grade sodium sulphate was 99.5% pure and contained sodium sulphate as the only source of sulphate ions. In order to check the supplier's claims the laboratory supervisor analysed a sample.

She dissolved 3.627 g of the sample in **a small volume of water (i)** and then added **an excess (ii)** of **a saturated solution of barium chloride (iii)**. After allowing the mixture to stand for several days, it was filtered and the residue was washed, with a **small volume (iv) of cold water (v)**, dried and weighed. The mass of dry solid was 5.644 g.

  a Explain, from a practical point of view, the significance of the five parts in bold type.
  b Can you suggest any ways in which the procedure could be modified in order to give more precise results?
  c What mass of residue would have been formed if the sample had been 99.5% pure, assuming that all of the SO_4^{2-} was precipitated and collected?
  d Do you think the supplier's claims were justified? Explain your answer.

Compounds of Group II elements (2)

Reactions with water and hydroxides

① If $M_{(s)}$ can be used to represent one mole of any Group II metal, write a general equation to show how these metals react with water.

② Which member of the Group does not react in this way?

③ How does the vigour of the reaction change as you go down the Group? Explain this trend.

④ This question is concerned with the solubilities of the Group II metal hydroxides.
a Copy and complete this table.

Element	Formula	Formula mass	Solubility (g/100g of water)	Solubility (mol dm^{-3})
Beryllium			insoluble	
Magnesium			0.00120	
Calcium			0.120	
Strontium			1.00	
Barium			4.70	

You may wish to use a spreadsheet to perform this calculation and to present your values in the form of a graph.

b How does solubility change as you go down the Group?

Period 3 (2)

> The electrical conductivity, melting points and boiling points of the elements depend largely on the type of bonding present.

① Explain how the bonding in the elements of Period 3 changes as the Period is crossed from left to right.

Conductivity

② State which of the elements of Period 3 conduct electricity and which do not.

③ Explain why some of the Period 3 elements can conduct electricity but others cannot.

Melting points and boiling points

④ The melting points (in Kelvin) of the elements of Period 3 are given in this table.

Element	Na	Mg	Al	Si	P	S	Cl	Ar
Melting point/K	371	923	933	1685	317	392	172	84

Plot these values on a graph.

⑤ Explain the variation in melting point across the Period.

⑥ The boiling points (in Kelvin) of the elements of Period 3 are given in this table.

Element	Na	Mg	Al	Si	P	S	Cl	Ar
Boiling point/K	1163	1378	2743	2632	533	718	239	87

Using a different vertical scale, plot these values on the same graph as the melting points – use a different colour for these points.

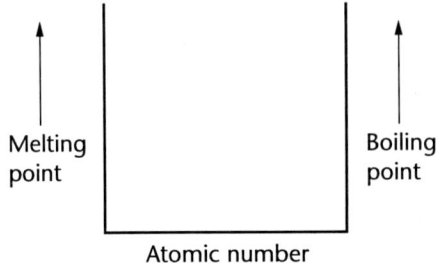

⑦ Why are the boiling points of the elements higher than their melting points?

⑧ Explain why silicon has the highest melting point but aluminium has the highest boiling point of the elements in Period 3.

Relative atomic and molecular mass

Relative atomic mass

Values of relative atomic masses for questions on this sheet:
C = 12, N = 14, O = 16, S = 32, Cl = 35.5, Br = 80.

① The following information was obtained from a mass spectrometer:

Ion	Mass/g
Fluorine $_9F^+$	3.154×10^{-23}
Carbon (isotope carbon-12) $_6C^+$	1.993×10^{-23}

Fluorine has one naturally occurring isotope. Calculate the relative atomic mass of fluorine.

② Calculate the relative atomic mass of each of the following elements.
   a One atom of silicon is 2.34 times as heavy as an atom of ^{12}C.
   b One atom of neodymium is 12.02 times as heavy as an atom of ^{12}C.

Relative molecular mass

③ Calculate the relative molecular mass of TCP (2,4,6-trichlorophenol) $C_6H_3O_1Cl_3$.

④ Transition metals form a family of compounds called carbonyls. The molecular formula of one such complex is $M_6(CO)_{16}$. The relative molecular mass of this compound is 1065.46. What is the relative atomic mass of M?

⑤ A double salt has the formula $XY(SO_4)_2.12H_2O$. X and Y are metals. The ratio of their relative atomic masses is 1:3.04. The relative molecular mass of this salt is 501. Calculate the relative atomic masses of X and Y. Use a Periodic Table to identify the metals.

Avogadro's constant and moles

> Avogadro's constant is 6.023×10^{23}

⑥ Describe the calculation you would use to find the mass of one mole of atoms from the mass of a single atom.

⑦ A drop of bromine (Br_2) has a volume of 0.05 cm^3. Calculate the number of molecules present in the drop. Density of $Br_2 = 3.12 \text{ g cm}^{-3}$.

> No. of moles = mass/A_r or M_r

> Use an appropriate number of significant figures in answers to calculations.
> Show full details of your method of calculation.

© Farrow R, Gibbens D, Stirrup M and Vowles R, 2000. AS Chemistry for AQA – Resource Pack, Heinemann

Module 1 — Atomic structure, bonding and periodicity — 4 Amount of substance

Avogadro's constant and moles (1)

Values of relative atomic masses for questions on this sheet:
H = 1, B = 11, C = 12, N = 14, O = 16, Na = 23, Al = 27, Cl = 35.5.

1. How many atoms of carbon are there in 1 mole of Bucky balls? Each ball is composed of 60 atoms of carbon.

2. A toxic metal was released into a river killing many fish. Analysis of a dead fish showed it to contain 0.2 ppm (by mass) of toxic metal in every 200 g of dead fish. The relative atomic mass of the toxic metal is 112. Calculate the number of atoms of toxic metal in 10 g of the dead fish.

3. Calculate the number of nitrate ions in 20 g of aluminium nitrate $Al(NO_3)_3 \cdot 9H_2O$.

4. Sodium metal is produced on a large scale by electrolysing molten sodium chloride. Electrolysis is carried out at 900 °C and atmospheric pressure.

   a Use the information below to calculate the mass of sodium produced when an electric current of 30 amps is passed through molten sodium chloride for 5 hours.

   b What volume of chlorine gas (in dm^3) is produced at the same time?

   (Cathode) $Na^+ + e^- \rightarrow Na$
   (Anode) $Cl^- \rightarrow \tfrac{1}{2}Cl_2 + e^-$

   > 1 mole of electrons = 96 487 coulombs of charge
   > A current of 1 ampere (amp) passing for 1 second carries 1 coulomb of charge
   > 1 mole of chlorine gas at atmospheric pressure and a temperature of 900 °C occupies 96 246 cm^3

5. Calculate the percentage by mass of boron in the compound $B_{10}C_2H_{12}$.

6. The density of benzene (C_6H_6) is 0.878 g cm^{-3}.

   a Calculate the volume occupied by 67 moles of benzene.

   b What is the mass of carbon in this volume of benzene?

7. One molecule of a white solid (at 0 °C) is composed of 3 atoms of hydrogen, 1 atom of nitrogen and 2.656×10^{-23} g of atoms of a third element. Calculate the M_r of the solid.

> Use an appropriate number of significant figures in answers to calculations.
> Show full details of your method of calculation.

Avogadro's constant and moles (2)

Values of relative atomic masses for questions on this sheet:
H = 1, C = 12, N = 14, O = 16, Mg = 24, Al = 27, P = 31, Cl = 35.5,
K = 39, Ca = 40, Fe = 56, Zn = 65, Ba = 137, Os = 190.

① Say whether each of the following statements is true or false.
  a 3 moles of zinc oxide have a mass of 306 g.
  b There are 2 moles of sodium ions in 80 g of sodium hydroxide crystals.
  c There are 0.25 moles of sulphur dioxide molecules in 16 g of sulphur dioxide.
  d There are 0.08 moles of oxygen atoms in 0.2 moles of sulphuric acid.
  e There are 1.2045×10^{24} potassium ions in 100 g of potassium hydrogencarbonate.
  f There are 15 moles of chloride ions in a mixture of 3 moles of magnesium chloride and 3 moles of iron(III) chloride.
  g 9.11 g of osmium contain the same number of atoms as there are in 3.6 g of phosphorus molecules (P_4).

② Duralumin is an alloy. The alloy contains 95% aluminium and 1% magnesium. The other component of the alloy is copper. How many moles of copper atoms are there in 34.6 g of duralumin?

Empirical and molecular formulae

> Before you start the questions below you need to be sure that you know what an empirical formula is and what a molecular formula is.

③ Find the empirical formula of a compound containing 81.07% barium and 18.93% oxygen.

④ a What is the simplest formula of a compound that is composed of 87.5% nitrogen and 12.5% hydrogen?
   b What additional information is required in order to calculate the molecular formula of the compound?

> Use an appropriate number of significant figures in answers to calculations.
> Show full details of your method of calculation.

Empirical and molecular formulae

Values of relative atomic masses for questions on this sheet:
H = 1, C = 12, N = 14, O = 16, Na = 23, Co = 59.

(1) A mass spectrometer determined the relative molecular mass of a hydrocarbon to be 128. Elemental analysis shows that the hydrocarbon contains 93.75% carbon. Calculate the empirical and molecular formula of the hydrocarbon.

(2) Analysis shows a compound to contain 36.364% oxygen and 54.545% carbon by mass. The remainder is hydrogen. If the compound is known to have a molecular mass in the range 70–100, determine the molecular formula of the compound.

(3) A red hydrated salt of a transition metal analyses as follows:
25.76% cobalt, 13.97% sulphur, 4.37% hydrogen and 55.90% oxygen.
  a Calculate the simplest formula of the hydrate.
  b How many moles of water are present in one mole of the hydrate?
  c The relative molecular mass of the hydrate is 229. What anion is present in the crystals of the hydrated metal salt?

(4) A drum containing baking powder was tested to find out what it contained. The majority of the material was a compound with the following composition:
carbon 14.29%, hydrogen 1.19%, oxygen 57.14% and sodium 27.38%.
  a Find the empirical formula of this compound.
  b Another compound in the mixture had the percentage composition: carbon 40.91%, hydrogen 4.55% and 54.54% oxygen. Find the empirical formula of this compound.
  c The molecular masses of the two compounds were 176 and 84, but not necessarily in that order. Find the molecular formula for each compound.

> Do not forget that the molecular formula is always a whole number times the empirical formula (which can be one).

Balancing equations

(5) State what is wrong in each of the equations.
  a $Ag^+(aq) + Zn(s) \rightarrow Zn^{2+}(aq) + Ag(s)$
  b $Rb(OH)_2(aq) + 2HBr(aq) \rightarrow RbBr_2(aq) + 2H_2O(l)$
  c $Mg(l) + H_2SO_4(aq) \rightarrow H_2(s) + MgSO_4(aq)$
  d $Cr_2(SO_4)_3(aq) + BaCl_2(aq) \rightarrow BaSO_4(s) + 2CrCl_3(aq)$

Module 1 — Atomic structure, bonding and periodicity — 4 Amount of substance

Balancing equations

① Copy out these equations and insert the missing species. Do not alter the number of reactant and product species shown in the equations.

  a $Cl^- + ? \rightarrow ICl_4^-$
  b $2IF_5 \rightarrow ? + IF_6^-$
  c $BeCl_2 + ? \rightarrow BeCl_4^{2-}$

② The following equations are incorrectly balanced. Rewrite the equations so they balance.

  a $3NH_3 + H_3PO_4 \rightarrow 3(NH_4)_3PO_4$
  b $Li_2CO_3 \rightarrow 2CO_2 + 2Li_2O$
  c $Cu^+ \rightarrow 2Cu^{2+} + Cu$
  d $P_4 + 5OH^- + H_2O \rightarrow PH_3 + 5H_2PO_2^-$
  e $3Fe^{2+} + Cl_2 \rightarrow 3Fe^{3+} + 2Cl^-$
  f $HgS + CaO \rightarrow 2Hg + CaSO_4 + 2CaS$
  g $ND_3 + 3CuO \rightarrow 3N_2 + 3Cu + D_2O$
  h $4CO + I_2O_5 \rightarrow 2I_2 + CO_2$

Constructing equations

③ Construct balanced chemical equations for the following reactions.
  a Strontium reacts with water to form hydrogen gas and strontium hydroxide.
  b At −60 °C krypton difluoride rapidly decomposes into its elements.
  c Aluminium carbide Al_4C_3 reacts with water to form methane gas and aluminium hydroxide.
  d Sucrose $C_{12}H_{22}O_{11}$ can be fermented to alcohol. In the overall equation for the fermentation process sucrose is shown to react with water to form ethanol and carbon dioxide gas.

④ V_2O_5 can be used to catalyse the gas phase reaction: $SO_2 + \tfrac{1}{2}O_2 \rightleftharpoons SO_3$
The catalyst works by using two reactions. Sulphur dioxide is oxidised to sulphur trioxide by vanadium pentoxide, which is reduced to vanadium tetraoxide V_2O_4. In the second reaction V_2O_4 is oxidised back to V_2O_5 by oxygen.
  a Construct balanced equations for these two reactions.
  b Combine the equations to show that V_2O_5 is unchanged at the end of the reactions and has acted as a catalyst.

Ionic equations

⑤ In each of the equations below identify the charge on the species in bold.
  a **C_2** $+ 2H_2O \rightarrow 2OH^- + C_2H_2$
  b $2UO^{2+} + 4H^+ \rightarrow$ **U** $+ UO^{2+} + 2H_2O$
  c $2IrCl_6^{2-} + H_2O \rightarrow$ **$2IrCl_6$** $+ \tfrac{1}{2}O_2 + 2H^+$
  d $[Be(H_2O)_4]^{2+} + 4OH^- \rightarrow$ **$[Be(OH)_4]$** $+ 4H_2O$
  e $4Pu^{3+} + 2HNO_2 + 4H^+ \rightarrow$ **4Pu** $+ N_2O + 3H_2O$

© Farrow R, Gibbens D, Stirrup M and Vowles R, 2000. AS Chemistry for AQA – Resource Pack, Heinemann

Reacting masses

Values of relative atomic masses for questions on this sheet:
H = 1, C = 12, N = 14, O = 16, Al = 27, S = 32, Cl = 35.5, K = 39, Cu = 63.5, Zn = 65.

1. 20 g of zinc sulphide (ZnS) are formed when 13.4 g of zinc metal (Zn) reacts with sulphur. What mass of sulphur combined with the zinc?

2. Carbon can be used to extract a sample of copper metal from copper(II) oxide.

    $C + 2CuO \rightarrow 2Cu + CO_2$

    Calculate the mass of copper that is made when 86.2 g of metal oxide are roasted with an excess of carbon (assume the reaction goes to completion).

3. An industrial plant in County Durham produces anhydrous aluminium(III) chloride. An excess of dry chlorine gas is used to convert aluminium metal into the chloride.

    $2Al + 3Cl_2 \rightarrow Al_2Cl_6$

    Calculate the theoretical maximum mass of aluminium chloride that could be formed when 60 slabs of aluminium are loaded into the reactor. Each slab weighs 2.75 kg. Temperature control is important. At very high temperatures small amounts of aluminium(I) chloride, AlCl, are formed together with the normal aluminium chloride. On cooling, the aluminium(I) chloride decomposes forming aluminium(III) chloride and aluminium metal. If 0.2 kg of aluminium(I) chloride is formed in the reactor, how much aluminium will contaminate the final quantity of aluminium chloride produced?

4. Amyl nitrite ($C_5H_{11}NO_2$) is a vasodilator. A research student found a method for making this compound on the Internet. *Some* of the details were downloaded and are given below.

    Dissolve $14\frac{1}{2}$ g of potassium nitrite (KNO_2) in 8 ml of water and in a separate vessel mix 15 g (18 ml) of amyl alcohol $C_5H_{11}OH$ with 30 g of concentrated sulphuric acid. Cool the latter and add it slowly to the solution of the nitrite with stirring.

    $C_5H_{11}OH + KNO_2 + H_2SO_4 \rightarrow C_5H_{11}NO_2 + H_2O + KHSO_4$

    a Which reactant or reactants is or are in excess?

    b If the student used the method above, calculate the maximum amount of amyl nitrite formed. Assume the reaction goes to completion.

    c The student successfully followed the method above and managed to make 12 g of the product. Calculate the percentage yield of amyl nitrite obtained in the preparation.

Reacting masses and volumes

Values of relative atomic masses for questions on this sheet:
H = 1, C = 12, N = 14, O = 16, S = 32, Cl = 35.5, Cu = 63.5.

Reacting masses

(1) The flowchart shows the stages involved in making a complex salt of copper.

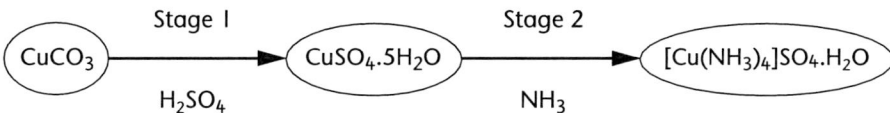

In Stage 1 an excess of copper carbonate is reacted with a known volume of 2M dilute sulphuric acid. The reaction produces an aqueous solution of copper sulphate. After removal of the excess copper carbonate the filtrate can be crystallised to produce crystals of copper sulphate ($CuSO_4.5H_2O$). In Stage 1, 24.95 g of crystals were made. The copper carbonate used was only 60% pure. (Assume all reactions go to completion. The reacting ratios are: Stage 1: 1 mol $CuCO_3$:1 mol H_2SO_4; Stage 2: 1 mol $CuSO_4.5H_2O$:1 mol $[Cu(NH_3)_4]SO_4.H_2O$.)

a Calculate the minimum amount of impure carbonate that must be used to react completely with 50 cm³ of 2M dilute sulphuric acid.

b Calculate the mass of complex salt $[Cu(NH_3)_4]SO_4.H_2O$ formed when 24.95 g of copper sulphate crystals are treated with an excess of aqueous ammonia.

c Calculate the minimum volume of 16M aqueous ammonia needed to be used in Stage 2.

Reacting volumes

(2) Use the ideal gas equation $PV = nRT$ to show that equal volumes of any two gases at the same temperature and pressure will contain the same number of moles.

(In each of the calculations that follow it is assumed that all the gases must be at the same temperature and pressure.)

(3) Calculate the volume of oxygen produced from 1 dm³ of nitrogen(I) oxide, in the reaction:

$2N_2O \rightarrow 2N_2 + O_2$

(4) If 60 cm³ of oxygen react with 30 cm³ of methane, will combustion be complete?

Reacting volumes

1. Air contains 20% by volume of oxygen. What volume of air is required to oxidise $2\,dm^3$ of ammonia to nitrogen(II) oxide?

   $4NH_3 + 5O_2 \rightarrow 6H_2O + 4NO$

2. Ammonia will burn in pure oxygen

   $4NH_3 + 3O_2 \rightarrow 2N_2 + 6H_2O$

   Some ammonia was burnt in pure oxygen and the total volume of the products was $132\,dm^3$. Calculate the volume of each reactant used.

3. Use the information that follows about a chemical reaction to construct a possible balanced equation for the reaction.

   $40\,m^3$ of chlorine gas reacts exactly with $40\,m^3$ of dinitrogen tetrafluoride, N_2F_4. $80\,m^3$ of product were formed in the reaction.

4. 25 volumes of a gaseous hydrocarbon required 100 volumes of oxygen gas for complete combustion. The products of the combustion are steam and carbon dioxide. Choose from this list the formula for the hydrocarbon:

   C_4H_{10}, C_3H_8, C_1H_4, C_5H_{12}, C_2H_6, C_6H_{14}

5. $500\,cm^3$ of chlorine and $600\,cm^3$ of hydrogen sulphide react to form hydrogen chloride and sulphur. The equation for the reaction is:

   $8H_2S(g) + 8Cl_2(g) \rightarrow 16HCl(g) + S_8(s)$

   What will be the volume of gas when the reaction is over?

Gas molar volume

> One mole of any gas occupies $22.4\,dm^3$ at $273\,K$ and $100\,kPa$.

6. Calculate the volume of hydrogen produced at stp when $3.0\,g$ of lithium react completely with an excess of water. The A_r of lithium is equal to 7.

7. What mass of potassium sulphite ($M_r = 158$) is needed to react with an excess of dilute nitric acid to produce $780\,cm^3$ of SO_2 gas (at stp)?

   $K_2SO_3 + 2HNO_3 \rightarrow 2KNO_3 + SO_2 + H_2O$

The ideal gas equation (1)

> You must be able to recall, rearrange and use $PV = nRT$
> $R = 8.31 \, \text{J} \, \text{mol}^{-1} \text{K}^{-1}$

① How many moles are there in $500 \, \text{m}^3$ of a sample of a gas at $20\,000 \, \text{Pa}$ and $400 \, \text{K}$?

② What is the volume of $7500 \, \text{g}$ of methane gas ($M_r = 16$) at a temperature of $385 \, \text{K}$ and a pressure of $159\,000 \, \text{Pa}$?

> $1 \, \text{cm}^3 = 1 \times 10^{-6} \, \text{m}^3$

③ What mass of nitrogen ($M_r = 28$) occupies $25.0 \, \text{dm}^3$ at $1.89 \times 10^5 \, \text{Pa}$ and $198 \, °\text{C}$?

④ Calculate the product PV for 1 mole of a gas at $401 \, \text{K}$. State the units.

⑤ What is the density of oxygen ($M_r = 32$), in $\text{g} \, \text{dm}^{-3}$, measured at $101 \, \text{kPa}$ and $273 \, \text{K}$?

⑥ A steel gas bottle contains $4500 \, \text{g}$ of oxygen. The pressure of the gas inside the bottle is $6492 \, \text{Pa}$. If $700 \, \text{g}$ of gas are slowly released from the bottle, what will be the new pressure of the remaining gas? (Assume that the temperature stays constant when the oxygen is released. $M_r(O_2) = 32$.)

The ideal gas equation in the determination of relative molecular mass

> The questions that follow include work covered in earlier sections of this unit.

⑦ $0.32 \, \text{g}$ of a solid compound is sublimed into a gas. The volume of gas formed, measured at $301 \, \text{K}$ and $106.5 \, \text{kPa}$ pressure, is $170.8 \, \text{cm}^3$. Calculate the relative molecular mass of the solid.

⑧ The vapour of a chloride of an element E occupies $82.07 \times 10^{-6} \, \text{m}^3$ at $227 \, °\text{C}$ and $101\,325 \, \text{Pa}$. The vapour had a mass of $379.4 \, \text{mg}$. This mass of vapour was made by treating the $95.8 \, \text{mg}$ of the element with an excess of chlorine gas. Calculate the molecular formula of the chloride.

> Be consistent with units.

The ideal gas equation (2)

Values of relative atomic masses for questions on this sheet:
C = 12, N = 14, O = 16, F = 19, Na = 23, Si = 28, Cl = 35.5, K = 39.

1 A liquid from an unlabelled aerosol was analysed and was found to be a single compound. 20 g of the liquid contains 6.28 g of fluorine, 11.74 g of chlorine and 1.98 g of carbon. In a separate investigation 0.951 g of a vaporised sample of the liquid occupied 300 cm^3 at 200 °C and a pressure of 103 kPa.

   a Use these results to calculate the relative molecular mass of the liquid.

   b Calculate the empirical and molecular formulae for this compound.

The ideal gas equation and reacting masses

2 The main component of dynamite is nitroglycerine, a liquid which has the formula $C_3H_5N_3O_9$. When nitroglycerine explodes the products of the reaction are all gases. The products are CO_2, H_2O, N_2 and O_2.

   a Write a balanced chemical equation for the explosion of nitroglycerine.

   > Hint: Revise the work you have done on balancing equations.

   b What is the total number of moles of gases produced when 10 cm^3 of nitroglycerine explode? (The density of nitroglycerine is 1.59 g cm^{-3}.)

   c The decomposition of nitroglycerine produces an enormous rise in pressure within milliseconds, which is the main cause of the explosion. If the reaction was so rapid that the products of the explosion occupied 3.00 dm^3 at 500 °C, what would the pressure be?

3 An air bag in a motor car is used to protect the driver in the event of a crash. When the car collides the bag quickly inflates. The chemicals NaN_3, KNO_3 and SiO_2 are stored in a canister under the dashboard of the car and reactions are initiated electrically at the moment of collision. The air bag is inflated with nitrogen gas produced by the reactions below. The equations are **not** balanced.

   $NaN_3 \rightarrow Na + N_2$ and $Na + KNO_3 \rightarrow K_2O + Na_2O + N_2$

   The oxides of sodium and potassium react further with SiO_2 to produce silicates.

   a Balance the equations.

   b What is the minimum mass of sodium azide, NaN_3, and potassium nitrate(V), KNO_3, needed to inflate a 100 dm^3 air bag, at 298 K and 101 kPa of pressure?

   > Be consistent with units.

Molarity and solutions

Values of relative atomic masses for questions on this sheet:
H = 1, Li = 7, C = 12, N = 14, O = 16, F = 19, Na = 23, Al = 27, S = 32, Cl = 35.5, Mn = 55, Cu = 63.5, Pb = 207.

Concentrations of solutions

1. What is the molarity of solutions with the following concentrations?
   a 5.4 g dm^{-3} LiNO$_3$
   b 0.97 g dm^{-3} MnSO$_4$
   c 0.35 g dm^{-3} CH$_3$COCH$_3$

2. What is the concentration of the each of the following in g dm^{-3}?
   a 1 M NaOH
   b 0.04 M CuCl$_2$
   c 0.001 M Al$_2$(SO$_4$)$_3$
   d 5 M HNO$_3$

3. Sodium fluoride has the formula NaF. Calculate its relative molecular mass. What mass of sodium fluoride would be required to make each of the following?
   a 3 dm^3 of 0.4 M solution
   b 10 cm^3 of 0.15 M solution
   c 250 cm^3 of 1.20 M solution

4. A chemist wants to prepare a standard solution which is 0.017 M with respect to NO$_3^-$ ions from lead nitrate Pb(NO$_3$)$_2$. There is only 0.68 g of crystals left in the store room. What total volume of solution must be prepared by the chemist for the solution to be 0.017 M?

5. An alcoholic drink is 40.0% ethanol by volume. Calculate the molarity of the drink with respect to ethanol. (Density of ethanol = 0.79 g cm^{-3}.)

Reacting volumes of solutions

6. Dilute hydrochloric acid is used for cleaning some metal surfaces. One bottle of cleaning solution contains 900 cm^3 of acid. Each 5 cm^3 of solution contains 0.365 g of HCl.
   a What volume of concentrated hydrochloric acid (12 M) must be diluted with water to provide enough cleaning solution to fill eight bottles?
   b How much water is used in the dilution of the concentrated acid?

7. Etching solution contains hydrofluoric acid (HF). A container of etching solution had drained onto the floor of its storage room. The spillage is normally dealt with by neutralising the acid with solid potassium hydrogencarbonate. Unfortunately there was only a concentrated solution of this compound available. The container of etching solution originally held 1.25 dm^3 of 0.72 M acid. 100 cm^3 of potassium hydrogencarbonate solution contained 21 g of compound. What volume of potassium hydrogencarbonate solution was needed to neutralise the acid?

© Farrow R, Gibbens D, Stirrup M and Vowles R, 2000. AS Chemistry for AQA – Resource Pack, Heinemann

Volumetric analysis (1)

Values of relative atomic masses for questions on this sheet:
H = 1, C = 12, O = 16, Na = 23, S = 32, Cl = 35.5, Rb = 85, Ba = 137.

① This is an extract taken from a chemical analysis book printed at the beginning of the last century. Read it carefully and then answer the questions below.

Standardisation of baryta solution: Pipette a 25 c.c. aliquot of baryta solution into a erlenmeyer flask. Add two drops of the thymol blue infusion and swirl. Dilute 25 c.c.s of 1 N muriatic acid to 250 c.c. of solution using a volumetric flask. Titrate the baryta solution to neutral point. Repeat the titration to achieve concordant titres. **Specimen results (titres of acid):** 24.40 c.c., 22.35 c.c., 22.35 c.c., 22.35 c.c.

a What is the thymol blue infusion acting as in the titration? What is the neutral point?

b The baryta solution contains $Ba(OH)_2$. The modern chemical name for muriatic acid is aqueous hydrogen chloride solution. A 1 N solution contains 36.5 g of compound in 1000 c.c.s of solution (1 c.c. = 1 cm³). Write an equation for the reaction occuring in the titration. Using a calculator find the molarity of the baryta solution and its concentration in $g\,dm^{-3}$.

② A powder toilet cleaner contains sodium hydrogensulphate $NaHSO_4$. A company which makes this toilet cleaner finds that one batch of sodium hydrogensulphate is contaminated with sodium sulphate. The batch is sent to the quality control laboratory for immediate analysis. The results of the analysis are shown below.

SAMPLE: contaminated batch 2/3241 **ANALYST:** G. B. Evers
METHOD OF ANALYSIS: Acidometric titration
25 cm³ portions of sample solution titrated with standard 0.100 M KOH
SAMPLE PREPARATION: 5.678 g/250 cm³ solution
RESULTS OF ANALYSIS: Mean titre 27.24 cm³ KOH(aq)

The acidity of the toilet cleaner is due only to the sodium hydrogensulphate. The reaction used in the acidometric titration is:

$$HSO_4^-(aq) + OH^-(aq) \rightarrow H_2O(l) + SO_4^{2-}(aq)$$

Calculate the percentage (by mass) of impurity in batch 2/3241.

③ 24.2 cm³ of 0.2 M hydrochloric acid was neutralised by 25 cm³ of an alkaline solution. The alkaline solution contained a 16.47 g mixture of sodium and rubidium hydroxide dissolved to give 1000 cm³ of solution. Calculate the mass of sodium hydroxide in the mixture.

Volumetric analysis (2)

Values of relative atomic masses for questions on this sheet:
H = 1, C = 12, N = 14, O =16, Cl = 35.5, Br = 80, Sr = 88.

(1) A bromide of an element was prepared and sealed in a glass ampoule. One ampoule contained 4.63 g of bromide. This ampoule was broken open under 250 cm³ of water (an excess). The bromide hydrolyses fully to form an insoluble element hydroxide and hydrobromic acid. 25 cm³ portions of the now acidic water were titrated with 0.2 M potassium hydroxide solution. The mean titre of alkali used was found to be 23.56 cm³. Determination of the M_r value of the bromide gave a value of 393. Using the titration data, construct a balanced equation for the hydrolysis of the bromide. Identify the element.

(2) A geology student wanted to know the percentage purity of a specimen of strontianite (strontium carbonate $SrCO_3$). The student's friend was studying A-level chemistry and offered to find the percentage purity of the specimen in one of the chemistry lessons. In the practical lesson 1.65 g of sample was treated with an excess of 0.200 M nitric acid. The volume of acid used was 100 cm³. The amount of unreacted nitric acid remaining was found by titration. The unreacted acid needed 23.6 cm³ of 0.100 M potassium hydroxide solution for neutralisation. Calculate the percentage by mass of strontium carbonate in the specimen of strontianite.

(3) The molecular formula of an inorganic acid salt is $KH_aC_4O_8.bH_2O$.
Use the information below from two separate experiments to find numerical values for a and b.

Experiment 1
25 cm³ portions of a 0.200 M solution of potassium hydroxide needed a mean titre of 33.30 cm³ of a 0.0500 M solution of the inorganic compound for complete neutralisation.

Experiment 2
10.980 g of the inorganic compound was gently heated until the mass no longer decreased and remained constant. The heat removed the water molecules from the compound. After heating the compound weighed 9.424 g.

Module 2 | Foundation physical and inorganic chemistry | 5 Energetics

Enthalpy change and calorimetry

32

① Write brief notes, with a definition if appropriate, on each of the following: (Refer to the student's book if you need some help.)
   a endothermic reaction
   b exothermic reaction
   c enthalpy change
   d standard enthalpy change
   e standard enthalpy change of formation
   f standard enthalpy change of combustion
   g heat change = $mc\Delta T$ (show that you understand specific heat and heat capacity)
   h Hess's law
   i mean bond enthalpy.

Standard enthalpy changes

② Write an equation, including state symbols, for the standard enthalpy of formation of:
   a $H_2O(l)$
   b $CH_3OH(l)$
   c $C_2H_5OH(l)$
   d $AlCl_3(s)$
   e $Na_2CO_3(s)$
   f $CuSO_4.5H_2O(s)$.

③ Write an equation, including state symbols, for the standard enthalpy of combustion of:
   a $CH_4(g)$
   b $C_4H_{10}(g)$
   c $CH_3OH(l)$
   d $C_2H_5OH(l)$
   e $CH_3COOH(l)$.

④ Calculate the standard enthalpy of combustion for ethane (C_2H_6), propane (C_3H_8) and pentane (C_5H_{12}) from the following information:
   a 5 g of ethane burned in excess oxygen produced 260 kJ
   b 5 g of propane burned in excess oxygen produced 252 kJ
   c 5 g of pentane burned in excess oxygen produced 244 kJ.

Calorimetry

Where required assume that 1 cm³ of an aqueous solution is 1 g.
The specific heat of an aqueous solution is 4.2 J g⁻¹ K⁻¹.
Assume that no heat is lost to, or gained from, the surroundings.

⑤ When 0.050 g sodium was added to 100 cm³ water the temperature rose by 1.24 °C.
   a Calculate the heat change in the reaction.
   b How many moles of sodium were used?
   c Calculate the enthalpy change for the reaction in kJ mol⁻¹.
   d Lithium undergoes a similar reaction with water. However, although it is more exothermic, the reaction is slower. Suggest why the experiment might be less accurate to measure the enthalpy change for the lithium–water reaction when carried out in practice.

⑥ When 100 cm³ 1 M sodium hydroxide were added to 100 cm³ 1 M hydrochloric acid the temperature rose by 6.8 °C.
   a Calculate the heat change in the reaction.
   b Convert your figure for the heat change into an enthalpy change in kJ mol⁻¹.

Module 2 | Foundation physical and inorganic chemistry | 5 Energetics

Calorimetry and using Hess's Law

Calorimetry

1 When 6.0 g of potassium nitrate (KNO_3) were dissolved in 100 cm^3 of water the temperature dropped by 2.1 °C.
  a Calculate the heat change in the reaction.
  b How many moles of potassium nitrate were used?
  c Calculate the enthalpy change for the reaction in kJ mol^{-1}.

2 When an excess of magnesium was added to 20 cm^3 of 0.2M copper sulphate solution the temperature of the solution increased from 18.3 °C to 43.0 °C.
  a Calculate the heat change in the reaction.
  b How many moles of copper sulphate were used?
  c Calculate the enthalpy change for the reaction in kJ mol^{-1}.

3 As an assessment exercise a student had to determine whether a given sample was methanol (CH_3OH) or ethanol (C_2H_5OH). Their standard enthalpies of combustion are −715 kJ mol^{-1} and −1371 kJ mol^{-1}, respectively.

The student burned 0.47 g of the sample and the heat produced increased the temperature of a calorimeter containing water from 19.0 °C to 53.3 °C. The total heat capacity of the calorimeter and water was 400 J K^{-1}. Calculate the heat change in the combustion reaction and use the data to deduce the identity of the sample.

4 A vessel and its contents have a heat capacity of 500 J K^{-1}. It is heated using a butane burner. (The standard enthalpy of combustion of butane is −2877 kJ mol^{-1}.)
  a Calculate the mass of butane (C_4H_{10}) needed to give a theoretical temperature rise of 30 °C.
  b Calculate the theoretical temperature rise occurring when 0.50 g of butane is burned.

Using Hess's law

5 Given:

$Mg(s) + Fe^{2+}(aq) \rightarrow Mg^{2+}(aq) + Fe(s)$ $\Delta H = -374$ kJ mol^{-1}
$Fe(s) + Cu^{2+}(aq) \rightarrow Fe^{2+}(aq) + Cu(s)$ $\Delta H = -154$ kJ mol^{-1}
$Zn(s) + Cu^{2+}(aq) \rightarrow Zn^{2+}(aq) + Cu(s)$ $\Delta H = -226$ kJ mol^{-1}

calculate the enthalpy changes for the following reactions:
  a $Mg(s) + Cu^{2+}(aq) \rightarrow Mg^{2+}(aq) + Cu(s)$
  b $Zn(s) + Fe^{2+}(aq) \rightarrow Zn^{2+}(aq) + Fe(s)$

6 Calculate the enthalpy change of the reaction:

$Cu(NO_3)_2(s) + 3H_2O(l) \rightarrow Cu(NO_3)_2.3H_2O(s)$

given these standard enthalpies of formation:

$Cu(NO_3)_2(s)$ −307 kJ mol^{-1}
$Cu(NO_3)_2.3H_2O(s)$ −1219 kJ mol^{-1}
$H_2O(l)$ −286 kJ mol^{-1}

© Farrow R, Gibbens D, Stirrup M and Vowles R, 2000. AS Chemistry for AQA − Resource Pack, Heinemann

Module 2 — Foundation physical and inorganic chemistry — 5 Energetics

Calculations using Hess's Law

34

① Calculate the enthalpy change of the reaction:

$$2Cu(NO_3)_2(s) \rightarrow 2CuO(s) + 4NO_2(g) + O_2(g)$$

given these standard enthalpies of formation:

Cu(NO$_3$)$_2$(s) −307 kJ mol^{-1}
CuO(s) −155 kJ mol^{-1}
NO$_2$(g) +34 kJ mol^{-1}

② a Calculate the missing ΔH values in the table opposite.

Use ΔH_f CO$_2$(g) = −394 kJ mol^{-1}
ΔH_f H$_2$O(l) = −286 kJ mol^{-1}

	ΔH_f/kJ mol^{-1}	ΔH_c/kJ mol^{-1}
CH$_4$(g)		−890
C$_2$H$_4$(g)		−1409
C$_4$H$_{10}$(g)	−125	
CH$_3$OH(l)		−715
C$_2$H$_5$OH(l)	−278	
CH$_3$COOH(l)		−876
CH$_3$COOC$_2$H$_5$(l)	−481	

b Use the data to calculate ΔH for the reaction:

$$CH_3COOH(l) + C_2H_5OH(l) \rightarrow CH_3COOC_2H_5(l) + H_2O(l)$$

Mean bond enthalpies

③ Calculate the enthalpy change of the reaction below by two different methods.

$$C_3H_8(g) + 5O_2(g) \rightarrow 3CO_2(g) + 4H_2O(g)$$

a Use the standard enthalpies of formation:

C$_3$H$_8$(g) −104 kJ mol^{-1}
CO$_2$(g) −394 kJ mol^{-1}
H$_2$O(g) −242 kJ mol^{-1}

b Use the following mean bond enthalpies quoted in kJ mol^{-1}.

C—C 348 C—H 412 O=O 496 C=O 804 O—H 463

c Account for any difference between the two values.

④ Use the mean bond enthalpies given below (in kJ mol^{-1}) to calculate the enthalpy change of each reaction.

a $N_2(g) + 3H_2(g) \rightarrow 2NH_3(g)$
b $H_2(g) + I_2(g) \rightarrow 2HI(g)$
c $H_2(g) + \frac{1}{2}O_2(g) \rightarrow H_2O(g)$

N≡N 944 H—H 436 N—H 388 I—I 151
H—I 299 O—H 463 O=O 496

⑤ The enthalpy change of the following reaction is −115 kJ mol^{-1}.

$$CH_4(g) + Cl_2(g) \rightarrow CH_3Cl(g) + HCl(g)$$

Calculate the mean C—Cl bond enthalpy, given the following values in kJ mol^{-1}.

C—H 412 Cl—Cl 242 H—Cl 431

© Farrow R, Gibbens D, Stirrup M and Vowles R, 2000. AS Chemistry for AQA – Resource Pack, Heinemann

Thermochemical calculations

(1) Potassium hydrogencarbonate decomposes according to the following equation:

$2KHCO_3 \rightarrow K_2CO_3 + CO_2 + H_2O$

Its enthalpy change can be calculated from the enthalpy changes for the following two reactions:

$K_2CO_3 + 2HCl \rightarrow 2KCl + CO_2 + H_2O$
$KHCO_3 + HCl \rightarrow KCl + CO_2 + H_2O$

A student obtained the following experimental results.

When 2.75 g of anhydrous potassium carbonate was added to 30 cm³ of 2M hydrochloric acid the temperature of the acid went up from 18.0 °C to 24.3 °C.

When 3.5 g anhydrous potassium hydrogencarbonate was added to 30 cm³ of 2M hydrochloric acid the temperature went down from 18.0 °C to 11.1 °C.

a Calculate the enthalpy change for each of the two reactions in kJ mol⁻¹.

b By considering the equations above, use your answers to part **a** to calculate the enthalpy change for the decomposition of potassium hydrogencarbonate.

c Suggest why the experiment would have been invalid if the student had used more than 4.14 g of potassium carbonate instead of 2.75 g.

(2) A student obtained the following experimental results:

a 5.0 g anhydrous copper sulphate was dissolved in 50 g water and the temperature of the water increased from 18.0°C to 28.8°C.

b 11.5 g of copper sulphate pentahydrate was dissolved in 50 g of water and the temperature of the water dropped from 18.0 °C to 16.9 °C.

  i Calculate the enthalpy change in kJ mol⁻¹ for each of the two processes above.
  ii The equations for the two processes could simply be written as:

   $CuSO_4(s) + 5H_2O(l) \rightarrow CuSO_4.5H_2O(aq)$
   $CuSO_4.5H_2O(s) \rightarrow CuSO_4.5H_2O(aq)$

   Use the results and the two equations to calculate the enthalpy change for the reaction:

   $CuSO_4(s) + 5H_2O(l) \rightarrow CuSO_4.5H_2O(s)$

(3) The standard enthalpies of formation of chromium(III) oxide and aluminium(III) oxide are –1128 kJ mol⁻¹ and –1669 kJ mol⁻¹, respectively.

a Calculate the standard enthalpy change for the reaction:

$2Al(s) + Cr_2O_3(s) \rightarrow Al_2O_3(s) + 2Cr(s)$

b How much heat would be produced if 18 g of aluminium were reacted with excess chromium(III) oxide under standard conditions?

c If the heat produced in part **b** were used to heat 1 kg of water at a temperature of 20.0 °C what would the final temperature of the water be, assuming no heat losses?

d Aluminium reacts similarly with iron(III) oxide whose standard enthalpy of formation is –822 kJ mol⁻¹.

  i Calculate the standard enthalpy change for the reaction between aluminium and iron(III) oxide.
  ii How much aluminium would be required to react with excess iron(III) oxide to increase the temperature of 1 kg of water from 20.0 °C to 100 °C?
  iii Would you expect the water to boil? Explain your answer.

© Farrow R, Gibbens D, Stirrup M and Vowles R, 2000. AS Chemistry for AQA – Resource Pack, Heinemann

Collision theory and rate of reaction (1)

Make sure that you know and understand collision theory.

Using collision theory

1. Account for each of the following by considering the energy and motion of the particles involved.
   a. A camp fire is initiated with newspaper and small twigs instead of large logs.
   b. A lump of bicarbonate of soda reacts more slowly with diluted vinegar than it does with neat vinegar.
   c. A decrease in pressure has virtually no effect on the speed of a reaction involving solids but lowers the speed of reactions involving gases.
   d. Hydrogen and nitrogen gas react faster with each other when iron powder is placed in the reaction mixture.
   e. Chemical change between solids is extremely slow.
   f. Magnesium ribbon reacts faster with hot water than with cold water.
   g. Reactions between gas molecules occur more slowly if the volume of the reaction vessel is increased.

2. We can express the rate of a reaction (number of fruitful collisions per cm^3 per second) as follows:

   rate of reaction = $O \times C \times E$

   O = orientation factor
   C = collision frequency
   E = energy factor

   a. In molecular terms what do O, C and E depend on?
   b. Which of these factors has the greatest influence on the rate of a reaction? Justify your answer.

Collision theory and rate of reaction (2)

① When thiosulphate ions ($S_2O_3^{2-}$) are mixed with an acidic solution colloidal sulphur is formed. The rate of the reaction can be studied using a colorimeter. The sulphur formed in the reaction scatters light and reduces the intensity of light that passes through the solution. A colorimeter can be used to measure the intensity of the light passing through a reaction mixture. A student used a colorimeter to investigate how the rate of the reaction varies with the concentration of thiosulphate ions. Different volumes of solution containing thiosulphate ions were made up to 10 cm³ with water. Each mixture in turn was placed in the colorimeter and reacted with 5 cm³ of acid solution (an excess). The time taken for the intensity of light passing through the mixture to fall to a reading of 50% was recorded. The colorimeter was maintained at a constant temperature throughout the investigation.

Some specimen results are shown below.

Volume of thiosulphate solution (cm³)	10	9	8	7	6	5	4	3
Time for light intensity to fall to 50% (seconds)	17	19	21	24	28	34	42	56

The reciprocal of the time taken for the intensity of light to fall to 50% of the initial intensity is a measure of the rate of reaction.

a Suggest another method for following the rate of the reaction.
b Why was each reaction mixture maintained at a constant temperature?
c The volume of thiosulphate solution was always diluted to the same volume. Why was the total volume maintained at a constant value?
d The student by accident had diluted the acid by a factor of 0.5 and then added 5 cm³ of this acid to 2 cm³ of thiosulphate solution diluted to 10 cm³. How would this mistake affect the rate of the reaction? Use collision theory to explain your answer.
e For each reaction mixture used in the experiment calculate the rate of reaction (1/time – in seconds⁻¹). Plot a graph of rate against volume of thiosulphate solution used. The volumes of thiosulphate solution are directly proportional to their concentrations.
f If 6.7 cm³ of thiosulphate solution are diluted to 10 cm³ with water, how long would it take for the reaction mixture to give a colorimeter reading of 50% of the initial intensity when the thiosulphate solution reacts with 5 cm³ of acid?
g Look at your graph. In words, describe the relationship between the rate of reaction and the concentration of thiosulphate ions. Give a simple mathematical equation to describe the relationship.
h Explain the shape of the graph in terms of the behaviour of the ions involved in the reaction
i On your graph sketch what you would expect to happen if the experiment was repeated at a higher temperature. Give reasons for your choice of sketch.

Factors influencing reaction rate (1)

This question involves some calculations.
Review your work on moles before attempting this question.
You will need a calculator.

① At room temperature and pressure a flask was connected to a gas syringe. 60 cm³ of 0.05 M dilute hydrobromic acid were placed in the flask. 2 g (an excess) of granules of a reactive metal were added, the flask was quickly stoppered and readings of the volume of gas in the syringe were recorded at quarter minute time intervals. The results of the experiment are shown below.

Time/ minutes	Volume of gas/ cm³	Time/ minutes	Volume of gas/ cm³
0	0	3	33
0.25	5	3.25	33.5
0.5	10	3.5	34
0.75	14	3.75	34.5
1	18	4	35
1.25	21	4.25	35.25
1.5	24	4.5	35.5
1.75	26	4.75	35.75
2	28	5	36
22.25	29.5	5.25	36
2.5	31	5.5	36
2.75	32	5.75	36

a A colourless gas is produced in the reaction. What is the gas?
b During the reaction the metal forms dipositive cations. Using the capital letter M to represent the chemical symbol of the metal, write a balanced equation for the reaction.
c The solution formed at the end of the reaction is coloured. In what block of the periodic table is the metal likely to be found?
d Suggest two other ways of following the rate of this reaction.

Factors influencing reaction rate (2)

Refer to the information on Worksheet 38 when answering the questions below. You will need a piece of graph paper for part **a**.

①
a Plot a graph of volume of gas produced against time.
b How long did it take for 29 cm^3 of gas to be produced?
c How long did it take for the reaction to be half completed?
d How much gas was made during the first, second and third minute of the reaction? What do these values indicate about how the rate of the reaction is changing with time? Use collision theory in your answer.
e On your graph sketch what you would expect to happen if the experiment was repeated using larger granules of excess metal. Label this A. Give reasons for your choice of sketch. The gas was still collected at room temperature and pressure.
f Sketch another line on your graph to show what you would expect to happen if the temperature was increased. Label this B. Again give reasons for your choice of sketch.
g The rate of the reaction doubles for every 10°C increase in temperature. By what factor would the reaction proceed more quickly at 80°C compared to 20°C? (It is not necessary to use the graph for this question.)
h What was the maximum volume of gas formed from the reaction at room temperature? What is the mass of this volume of gas?
i What is the average initial rate of formation of gas in the first half minute of the reaction in units of:
  i cm^3 of gas formed s^{-1}
  ii mol of gas formed s^{-1}?

> 1 mole of gas at room temperature and pressure occupies 24 dm^3.

j Work out the average initial rate of consumption of hydrobromic acid in mol dm^{-3} s^{-1} over the first half a minute of the reaction.

> Remember to use the mole ratio from the balanced chemical equation.

k For the reaction being investigated it is found that the average initial rate of formation of gas is directly proportional to the concentration of the acid in mol dm^{-3}. Calculate the average initial rate (in mol gas s^{-1}) for the first half minute of the reaction using:
  i 0.025 M HBr
  ii 5 cm^3 of 1.2 M hydrobromic acid diluted to give 60 cm^3 of acid, both reacting with an excess of metal granules at room temperature and pressure.
l The reaction of granules of the reactive metal is much slower with 0.05 M ethanoic acid than 0.05 M HBr at room temperature. Account for the difference in the rate of reaction.

Maxwell–Boltzmann distribution

1. The diagram shows the Maxwell–Boltzmann distribution of molecular energies at a temperature t_a for a mixture of gases which react with each other.

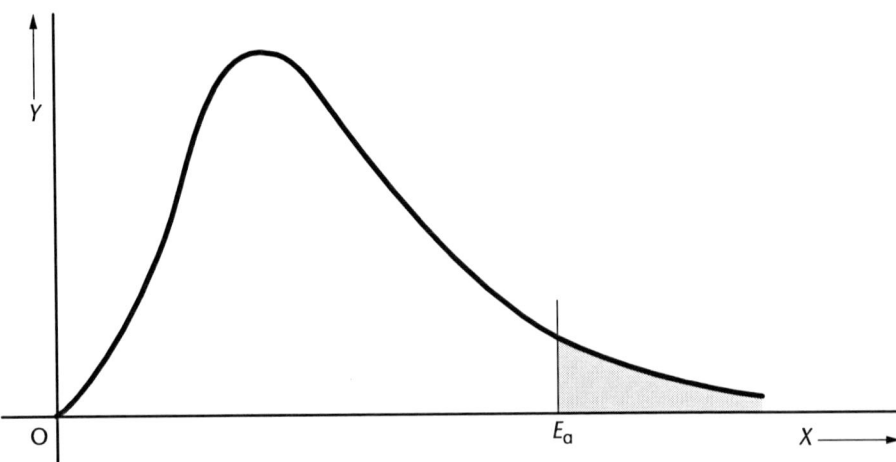

a The axes are labelled X and Y. What do X and Y represent?
b The reaction between the gases is not instantaneous. Explain this.
c What does the unshaded area under the curve represent?
d Copy the diagram and label the horizontal and vertical axes. On your diagram sketch a second distribution curve for the same mixture of gases at a lower temperature than t_a. Label the new distribution t_b. Use the distribution curves to explain why the reaction is slower at t_b.
e Are the areas under the two curves equal? Explain.
f What change to the mixture of gases, other than a change in temperature, would cause the unshaded area to alter?
g Would the distribution of molecular energies change if a smaller mass of gases were used at the temperature t_a? Explain your answer.

Reaction profiles

2. Using graph paper, draw to scale a simple energy profile to represent the following reaction:

$$H_2O_2(aq) \rightarrow H_2O(l) + \tfrac{1}{2}O_2(g) \qquad \Delta H = -97.7 \text{ kJ mol}^{-1}$$

The activation energy for this reaction is 75.3 kJ mol^{-1}.

a Label your profile.
b If the reaction was reversible, calculate the activation energy for the back reaction. Show the activation energy for the back reaction on your profile.
c Colloidal platinum reduces the activation energy for the forward reaction by 26.4 kJ mol^{-1}. On your diagram draw a possible energy profile for the catalysed reaction.

Catalysts

① Find out about the action, types and uses of catalysts from your school library or any other suitable sources of information.

② *"Catalysts have contributed to an improved standard of living."*
Discuss this statement. Write a short essay using this as the title.

③ In what part of the Periodic Table are elements and their compounds likely to be found which act as good catalysts?

④ Use the collision theory to explain the effect that a homogeneous catalyst has on the rate of a chemical reaction. Give an example of such a catalyst and a reaction it catalyses.

⑤ What is the difference between a homogeneous and heterogeneous catalyst in terms of the physical state and mode of action of the catalyst?

⑥ Nature's catalysts are mainly enzymes. How do enzymes differ from most inorganic catalysts?

⑦   'Green chemistry 2000 and beyond – environmentally friendly catalysts'

Using the title above, prepare a presentation. You could use one example and do a solo presentation or form a team with other students and present several examples.
You will need to do some research. You must plan and use different sources to search for and select information. As part of your presentation you must produce a small hand-out for members of your audience. The hand-out will allow the listeners to take an active part in your presentation. Your hand-out must contain different styles and methods of presenting information including the use of numerical, pictoral, graphical and photographic images.

You can use a variety of techniques to put forward your presentation.

Getting started ...
There is an interesting article by James Clark in the publication *Chemistry Review*, Vol. 2, No. 5, May 1993, pp. 26–31 (published by Philip Allan) which describes new ways of carrying out the Friedel–Crafts reaction using environmentally friendly catalysts.

Module 2 | Foundation physical and inorganic chemistry | 7 Equilibria

Establishing equilibria

Remember that the word reactant applies to all of the substances on both sides of the equilibrium equation.

1. In the equation $N_2(g) + 3H_2(g) \rightleftharpoons 2NH_3(g)$ what does the $\rightleftharpoons$ sign represent?

2. We usually abbreviate the term "dynamic equilibrium" to just "equilibrium". Explain the terms dynamic and equilibrium in this context.

3. Once equilibrium has been established what do you know about:
   a the rates of the forward and reverse reactions and
   b the amounts of reactants present?

4. What would you expect to happen to the forward and reverse reaction rates if some of the ammonia was removed? Would the system still be in equilibrium? Explain your answer.

5. Nitrogen and hydrogen were mixed and allowed to react. The rate of the forward reaction was measured at regular time intervals. The following results were obtained.

Time/min	0	2	4	6	8	10	12
Rate of reaction/ mol dm^{-3} s^{-1}	2.400	1.600	1.152	0.864	0.720	0.672	0.672

   a Plot these data on a graph with time on the horizontal axis.
   b At what time was equilibrium established?
   c Sketch on your graph how you think the rate of the reverse reaction would have changed over the same time period.

   Hint – at the start there was no $NH_3(g)$.

6. Write an equation, including state symbols, for the reaction between dilute hydrochloric acid and marble chips (calcium carbonate). Explain why this reaction goes to completion, in an open vessel, rather than reaching a position of equilibrium.

Applying Le Chatelier's Principle

> Make sure that you know Le Chatelier's Principle and how to apply it.

Qualitative effects of changing the temperature

1 Dinitrogen tetroxide, N_2O_4, is a dark brown gas that dissociates, at temperatures above 700 K, to form nitrogen dioxide, NO_2, which is pale yellow. Molecules of NO_2 can associate to form N_2O_4. The equilibrium can be represented by the equation:

$$N_2O_4(g) \rightleftharpoons 2NO_2(g)$$

When the temperature of the mixture is increased from 700 K to 900 K its colour fades. Explain this finding.

2 What would you expect to happen to the colour of the mixture if the temperature was kept constant, at 800 K, but the pressure was increased? Explain your answer.

Establishing equilibrium

3 One mole of dinitrogen tetroxide was put into a reaction vessel at 900 K and the mixture was allowed to react. The data in the table refer to the number of moles of $N_2O_4(g)$ present at 5-minute intervals after the start of the reaction.

Time/min	0	5	10	15	20	25	30
No. of mol of N_2O_4	1.00	0.66	0.48	0.36	0.30	0.28	0.28

a Plot these values on a graph with time on the horizontal axis.
b Add to your graph the change in the number of moles of NO_2 over the same time.

> Hint – you can see how many moles of N_2O_4 have reacted.

c How many moles of N_2O_4 would you expect to be present after 60 minutes? Explain your answer.

4 How would you expect a system of $N_2O_4(g)$ and $NO_2(g)$, in equilibrium, to respond to the addition of a small amount of oxygen?

A gaseous equilibrium

Changes in temperature and pressure

Make sure that you know Le Chatelier's Principle.

1. The percentage yield of product obtained for a gaseous reaction varies with the temperature and pressure. The yield was measured at a range of pressures at two different temperatures. The results were plotted to give the graphs shown.

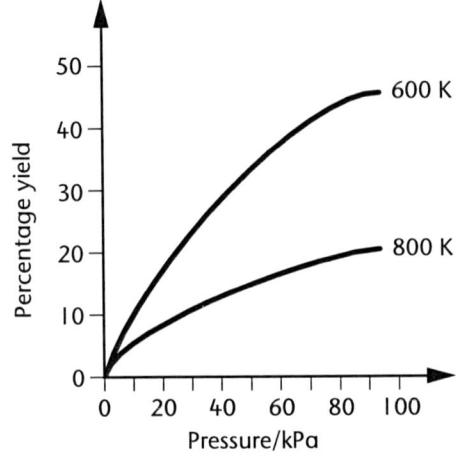

   a Is the reaction exothermic or endothermic? Say how you made your decision.

   b Does the number of moles of gas increase, decrease or stay the same as the reaction proceeds? Explain how you made your decision.

   c At 400 K and 50 kPa the percentage yield is 54%. What would you expect to happen to the yield if:
      i   the temperature was decreased
      ii   the pressure was increased
      iii a catalyst was added?

2. The reaction $R(g) + T(g) \rightleftharpoons W(g)$ is exothermic. State, qualitatively, the conditions you would use in order to obtain as high a yield of W in as short a time as possible and explain your reasoning.

Production of ethanol

> Ethanol can be made by the direct hydration of ethene (obtained in large amounts from the cracking of petroleum) or by the fermentation of sugars. In Britain the preferred industrial route for making ethanol for the chemical industry is direct hydration. The chemical reaction is:
>
> $$CH_2=CH_2(g) + H_2O(g) \rightleftharpoons CH_3CH_2OH(g) \quad \Delta H = -71.1 \, kJ\,mol^{-1}$$
>
> The conditions used are 600 K and 6000 kPa with a phosphoric acid catalyst on a solid inert support.

1. Why is a pressure of 6000 kPa used?

2. Why is the pressure not higher?

3. What would you expect to happen to the yield if the temperature was increased?

4. Considering your answer to question 3, say why the temperature is not lowered below 600 K.

5. Why is a catalyst particularly important in this reaction?

6. Explain why the catalyst is supported on an inert solid.

7. Under the conditions stated, the yield of ethanol is about 4%. How is it possible for an industrial process to be economic with such a low yield?

8. Direct hydration is a continuous process capable of producing large amounts of ethanol per hour.
   a What are the advantages of a continuous process compared to a batch process?
   b Why is it desirable to produce large amounts of ethanol quickly?

9. Alcoholic drinks are produced by fermentation, which is a batch process. Why is this so?

Production of ammonia

① What are the raw materials used in the manufacture of ammonia by the Haber Process?

> The equations for two of the reactions involved in the production of ammonia in the Haber Process are:
>
> $CH_4(g) + H_2O(g) \rightleftharpoons CO(g) + 3H_2(g)$ (1)
> $CO(g) + H_2O(g) \rightleftharpoons CO_2(g) + H_2(g)$ (2)
>
> Both of the reactions are exothermic.

② Say what you would expect to happen to the equilibrium position of each reaction if:
   a the temperature was decreased
   b the pressure was decreased.

③ Use your answers to question 2 to decide what are the best **qualitative** conditions for each reaction.

④ The equilibrium position gives an indication of the yield (the percentage of reactants that is converted to products). This is an important consideration when deciding the reaction conditions, but so are reaction rates and running costs. Look at your answer to question 3 and identify the problems associated with using the optimum equilibrium conditions.

> The reaction in which ammonia is produced is:
>
> $N_2(g) + 3H_2(g) \rightleftharpoons 2NH_3(g)$ $\Delta H = -92\,kJ\,mol^{-1}$

⑤ What are the optimum conditions to ensure a good yield of ammonia?

⑥ What are the conditions that are actually used for this reaction?

⑦ Why are the actual conditions said to be a compromise and why is the catalyst so important?

⑧ What is most of the ammonia used for?

Production of sulphuric acid

Sulphuric acid is a versatile reagent with many uses, making it one of the most important industrial chemicals. You will come across it being used as a strong acid, catalyst, dehydrating agent (and drying agent) and oxidizing agent. It is also a sulphonating agent. Some sulphuric acid is still produced by the Lead Chamber Process but modern plants use the Contact Process.

① A simplified scheme of the reactions involved in the Contact Process is shown. Copy it out and fill in the blank spaces.

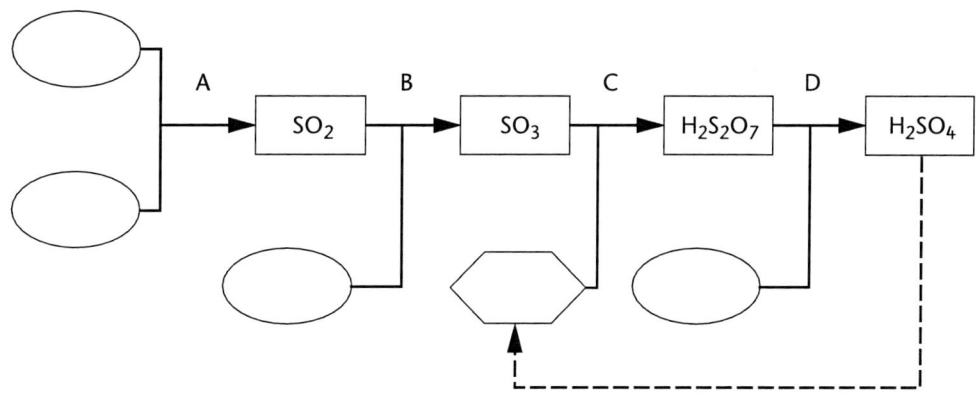

② What do the ovals and boxes represent?

③ Write out equations for stages A and B.

④ The ΔH values for these two reactions are $-297\,\text{kJ}\,\text{mol}^{-1}$ and $-192\,\text{kJ}\,\text{mol}^{-1}$ respectively. From what you know about equilibrium theory, say what the optimum conditions would be for these reactions, assuming that all of the reactants are gases. (Liquid sulphur is sprayed into the reaction chamber.)

⑤ In practice, stages B and C are combined in a number of stages so that 99.5% of the SO_2 is converted to SO_3 at a pressure below 1000 kPa. Explain why the pressure is not increased above 1000 kPa.

⑥ Why is it important to ensure that there are no leaks from the plant from an environmental point of view?

⑦ If you were asked to decide where to build a new Contact Process plant in Britain what factors would you take into account and where would you build it?

Oxidation and reduction

1 When lead(IV) oxide reacts with carbon, both oxidation and reduction occur:

$$PbO_2(s) + C(s) \rightarrow Pb(l) + CO_2(g)$$

with reduction from PbO_2 to Pb, and oxidation from C to CO_2.

  a If oxidation means 'adding oxygen', what does reduction mean in this simple context?
  b What is the reducing agent in this reaction?
  c What do the terms s, l and g mean?

2 Copy the following reactions and add oxidation and reduction arrows:
  a $CuO(s) + H_2(g) \rightarrow Cu(s) + H_2O(g)$
  b $Fe_2O_3(s) + 2Al(s) \rightarrow 2Fe(l) + Al_2O_3(s)$

Redox and electrons

3 Describe the following reactions, shown as half-equations, in words and use the OilRig mnemonic to work out what is happening in Redox terms.
  a $2K \rightarrow 2K^+ + 2e^-$ and $\frac{1}{2}O_2 + 2e^- \rightarrow O^{2-}$
  b $Na \rightarrow Na^+ + e^-$ and $\frac{1}{2}Cl_2 + e^- \rightarrow Cl^-$
  c $Fe \rightarrow Fe^{2+} + 2e^-$ and $S + 2e^- \rightarrow S^{2-}$
  d $2Fe \rightarrow 2Fe^{3+} + 6e^-$ and $1\frac{1}{2}O_2 + 6e^- \rightarrow 3O^{2-}$

Displacement

4 Convert these displacement reactions into their ionic form and use this to show where oxidation and reduction are taking place (ignoring the complex anions).
  a $Mg(s) + FeSO_4(aq) \rightarrow Fe(s) + MgSO_4(aq)$
  b $Ni(s) + Pb(NO_3)_2(aq) \rightarrow Pb(s) + Ni(NO_3)_2(aq)$
  c $Zn(s) + CuSO_4(aq) \rightarrow Cu(s) + ZnSO_4(aq)$
  d $CaBr_2(aq) + Cl_2(g) \rightarrow CaCl_2(aq) + Br_2(aq)$
  e $2KI(aq) + Cl_2(aq) \rightarrow 2KCl(aq) + I_2(aq)$
  f $Mg(s) + 2HCl(aq) \rightarrow MgCl_2(aq) + H_2(g)$
  g $2Na(s) + 2H_2O(l) \rightarrow 2NaOH(aq) + H_2(g)$

Oxidation states (1)

① State three guidelines that are used to work out oxidation states.

② What is the oxidation state for each of the following ions/atoms?
  a Al^{3+}
  b S^{2-}
  c Mg
  d I^-
  e Cu^{2+}
  f Cu^+
  g He

③ For a compound, the sum of oxidation states is zero. If chloride ions are Cl^-, what is the oxidation state of the metal in each of the following compounds?
  a $TiCl_4$
  b $CuCl_2$
  c $AlCl_3$
  d $RbCl$

④ To work out oxidation states, covalent molecules are treated as if they were ionic. If hydrogen is taken as +1, what is the oxidation state of the other element in each of the following simple compounds?
  a HCl
  b PH_3
  c CH_4
  d H_2O
  e H_2S
  f NH_3

⑤ Which of the following elements have just one common oxidation state in compounds and what is it in each case?

  Metals: K   Ca   Fe   Cr   Pb   Na   Al
  Non-metals: Cl   F   Br   S   N

⑥ Using your knowledge of elements with just one common oxidation state, copy and complete the following table:

Compound	Element with just one common state	Common oxidation state	Second element	Variable oxidation state
Li_2S				
$FeBr_3$				
PCl_3				
PCl_5				
CrO_3				
Cr_2O_3				
ClO_2				
CaC_2				

⑦ What is the oxidation number of:
  a hydrogen in lithium hydride, LiH
  b oxygen in hydrogen peroxide, H_2O_2?

Oxidation states (2)

1. Calculate the oxidation number of phosphorus in sodium phosphate, Na_3PO_4, given that:

Fixed oxidation state of Na = +1   Fixed oxidation state of O = –2

Number of Na atoms = 3   Number of O atoms = 4

Total = 3 × +1 = +3   Total = 4 × –2 = –8

Given that the compound number is 0, the oxidation number of P here must be ...?

2. Use the method above to find the oxidation number of:
   a. sulphur in magnesium sulphate, $MgSO_4$
   b. iodine in potassium iodate, KIO_3
   c. nitrogen in calcium nitrate, $Ca(NO_3)_2$
   d. chromium in potassium chromate, K_2CrO_4
   e. chromium in potassium dichromate, $K_2Cr_2O_7$

3. What is the maximum oxidation state of:
   a. C   b. P   c. S   d. Br

What is this maximum value equal to in each case?

4. What is the sum of the oxidation states for each of the molecular ions shown?
   a. BrO_4^- b. HCO_3^- c. HPO_4^{2-} d. N^{3-}

5. Find the oxidation number for:
   a. Cl in ClO_4^-
   b. P in $H_2PO_4^-$
   c. B in $B_4O_7^{2-}$

What's in a name?

(1) At AS level, elements with variable oxidation states must have these shown as Roman numerals in their names. For example: Fe_2O_3 would be named iron(III) oxide.
Give the full name for each of the following compounds:
  a $CuSO_4$
  b PbS
  c MnO_2
  d $Cr_2(SO_4)_3$
  e $FeSO_4$

(2) In very old text books, you will come across compounds called ferrous chloride, $FeCl_2$, and ferric chloride, $FeCl_3$. What would these compounds be called today?

(3) Use the oxidation numbers to help you work out the chemical formulae of:
  a cobalt(II) chloride
  b silver(I) nitrate
  c zinc(II) carbonate
  d lead(II) oxide
  e lead(IV) oxide
  f antimony(V) chloride

(4) For elements with variable oxidation states in anions, the Roman numerals come after the second part of the compound name.
What is the oxidation state of:
  a nitrogen in potassium nitrate(III), KNO_2
  b manganese in potassium manganate(VII), $KMnO_4$
  c chlorine in lithium chlorate(V), $LiClO_3$

(5) Explain the difference between sodium chloride and sodium chlorate(I). Give their respective formulae.

(6) Copy and complete the table below.

Old-fashioned ('trivial') name	Formula	Modern name
sodium sulphite		
potassium chlorite		
calcium nitrite		
ferrous hydroxide		

© Farrow R, Gibbens D, Stirrup M and Vowles R, 2000. AS Chemistry for AQA – Resource Pack, Heinemann

Oxidation states and Redox

1 Lead(IV) oxide is reduced to lead by carbon, which is in turn oxidised to carbon dioxide:

$$PbO_2 + C \rightarrow Pb + CO_2$$

Oxidation numbers: +4   0   0   +4

a Does the oxidation number of lead increase or decrease in this reaction?
b Does the oxidation number of carbon increase or decrease?
c What is the rule that links the change in oxidation number to oxidation and reduction?

2 Zinc will displace copper from copper sulphate solution:

$$Zn + CuSO_4 \rightarrow ZnSO_4 + Cu$$

a Work out the oxidation number changes for Zn and Cu in this reaction and decide which is being oxidised and which reduced.

$$Zn + Cu^{2+} \rightarrow Zn^{2+} + Cu$$

b Look at the same reaction in its ionic form and use the OilRig method to work out which is oxidised and which reduced. Do the two methods give the same answer?

3 Bromine displaces iodine from potassium iodide. Write full and ionic equations for this reaction and use the two methods to work out which is being oxidised and which reduced (as in question 2). Again, compare your answers.

4 In the following ionic equation, use the oxidation state method to find out whether the sulphur and iodine are being oxidised or reduced.

$$2IO_3^- + 5HSO_3^- \rightarrow I_2 + 5SO_4^{2-} + 3H^+ + H_2O$$

5 In the following ionic equation, use the oxidation state method to find out whether the sulphur and bromine are being oxidised or reduced.

$$SO_2 + Br_2 + 2H_2O \rightarrow 4H^+ + SO_4^{2-} + 2Br^-$$

Oxidising and reducing agents

6 Fe^{3+} ions are changed to Fe^{2+} ions by iodide ions. Is this oxidation or reduction? (OilRig!)

Use the oxidation state method to confirm this. What is happening to the iodine?

$$Fe^{3+} + I^- \rightarrow Fe^{2+} + \tfrac{1}{2}I_2$$

Iodide is acting as a **reducing agent** here. Explain what this means and link it to the loss or gain of electrons.

7 Potassium dichromate is an **oxidising agent** that was used in early breathalysers, because it changed colour from orange to green as it oxidised any alcohol in the breath.

$$Cr_2O_7^{2-} \rightarrow 2Cr^{3+}$$

Orange → green

Work out the oxidation numbers for chromium in this reaction and state what is happening to it in terms of oxidation or reduction. Are electrons gained or lost?

Simple half-equations

1 State whether each of the following half-equations shows oxidation or reduction.
  a $Cu^+ \rightarrow Cu^{2+} + e^-$
  b $S^{2-} \rightarrow S + 2e^-$
  c $Li^+ + e^- \rightarrow Li$
  d $O_2 + 4e^- \rightarrow 2O^{2-}$

2 Write half-equations for:
  a the reduction of a lead(IV) ion to a lead(II) ion
  b the oxidation of an barium atom to a barium ion
  c the oxidation of two fluoride ions to one fluorine molecule
  d the reduction of an iodine molecule to two iodide ions.

More complex half-equations

3 Copy the table below and use it to help you work out and balance the half-equation for the oxidation of hydrogen iodide to iodine.

Write down the formulae and balance the atoms undergoing redox	[HI $\rightarrow$ I_2] unbalanced...
Balance any oxygen atoms present by adding water	Not required
Balance any hydrogen atoms by adding H^+ ions	
Balance the charges by adding electrons	

4 Copy the table below and use it to help you work out and balance the half-equation for the reduction of manganate(VII) ions to manganese(II) ions.

Write down the formulae and balance the atoms undergoing redox	$MnO_4^- \rightarrow Mn^{2+}$
Balance any oxygen atoms present by adding water	
Balance any hydrogen atoms by adding H^+ ions	
Balance the charges by adding electrons	
Check that the number of atoms is balanced, the charges are balanced, and that the number of electrons transferred is equal to the change in oxidation number	

5 Use a similar table to help you work out and balance the half-equation for the reduction of chlorate(I) ions to chloride ions in bleach:
  $ClO^- \rightarrow Cl^-$

Constructing whole equations

1 Copy this table and complete it to show the full balanced equation from two half-equations for the reaction of bleach with iodide ions.

The half-equation for the oxidation part	$ClO^- + 2H^+ + 2e^- \rightarrow Cl^- + H_2O$
The half-equation for the reduction part	$2I^- \rightarrow I_2 + 2e^-$
Multiply up to match electrons (if necessary)	
Add the two together	
Cancel out the electrons	
Check that atoms and charges balance, add state symbols	

2 Copy this table and complete it to show the full balanced equation from two half-equations for the reaction of iron(II) ions with manganate(VII) ions.

The half-equation for the oxidation part	$Fe^{2+} \rightarrow Fe^{3+} + e^-$
The half-equation for the reduction part	$MnO_4^- + 8H^+ + 5e^- \rightarrow Mn^{2+} + 4H_2O$
Multiply up to match electrons	
Add the two together	
Cancel out the electrons	
Check that atoms and charges balance, add state symbols	

3 Copy this table and complete it to show the full balanced equation from two half-equations for the reaction of hydrobromic acid and sulphuric acid.

The half-equation for the oxidation part	$Br^- \rightarrow \frac{1}{2}Br_2 + e^-$
The half-equation for the reduction part	$SO_4^{2-} + 4H^+ + 2e^- \rightarrow SO_2 + 2H_2O$
Multiply up to match electrons	
Add the two together	
Cancel out the electrons	
Check that atoms and charges balance, add state symbols	

Module 2 — Foundation physical and inorganic chemistry — 9 Group VII, the halogens

Trends

① **a** Use the data below to plot graphs of atomic radius, boiling point and electronegativity each against atomic number for Group VII elements. Use a spreadsheet to help you do this.

Group VII element	Period number	Atomic number	Atomic radius/nm	Boiling point /°C	Electro-negativity
F	2	9	0.071	−188	4.0
Cl	3	17	0.099	−35	3.0
Br	4	35	0.114	59	2.8
I	5	53	0.133	184	2.5

b For each graph, describe the trend and try to explain it in terms of the atomic structure of these atoms.

c You will notice a pronounced bend in the trendlines around chlorine. Why is it that the F–Cl section is so different from the Cl–Br and Br–I sections in each case? Think about the change in electronic structure from period to period down the group.

② A full set of data for astatine is not available as this is an artificial radioactive element with a very short half-life.

a Extrapolate from your graphs above, to predict its atomic radius, boiling point and electronegativity.

b Which of these figures do you think is likely to be the most reliable?

c The atomic radii values of Ne, Ar, Kr and Xe are 0.065, 0.095, 0.11 and 0.13 nm respectively. How do these values compare to those of F, Cl, Br and I?

d Find out the atomic radius value for radon and compare it to your prediction for astatine from part **a**. How reliable do you think your estimate is now?

e The boiling point of astatine is 610 K. Convert this to °C and check your prediction from part **a**.

③ The halogens are all oxidising agents as they are electron acceptors, forming single-charged anions.

For example: $Br_2 + 2e^- \rightarrow 2Br^-$

Use your atomic radius graph to predict which halogen will be the strongest oxidising agent and which the weakest. Explain your answer.

© Farrow R, Gibbens D, Stirrup M and Vowles R, 2000. AS Chemistry for AQA – Resource Pack, Heinemann

Halogens are oxidising agents

1 Halogens are oxidising agents. That is, they are electron acceptors. For example, when chlorine reacts with sodium to make common salt the equations are:

Half-equations: $2Na \rightarrow 2Na^+ + 2e^-$ sodium loses electrons (OIL – oxidation)
  $Cl_2 + 2e^- \rightarrow 2Cl^-$ chlorine accepts electrons (RIG – reduction)
Added together: $Cl_2 + 2e^- + 2Na \rightarrow 2Na^+ + 2e^- + 2Cl^-$
Cancelling the $2e^-$: $Cl_2 + 2Na \rightarrow 2Na^+ + 2Cl^-$

Use the method above to show how oxidation and reduction occur in the following halogen reactions:

  a Iodine reacts with potassium to form potassium iodide, KI
  b Fluorine reacts with magnesium to form magnesium fluoride, MgF_2
  c Bromine reacts with iron(II) ions to give iron(III) ions.

2 **a** What is the trend in oxidising ability of the halogens in terms of their position in Group VII?
  b Write the half-equations for the reaction between chlorine water and potassium iodide solution.
  c State where oxidation and reduction are taking place.

Displacement reactions

3 The halogens show their elemental colours when dissolved in a solvent such as tetrachloroethene (dry cleaning fluid). If chlorine dissolved in tetrachloroethene (a green solution) is carefully mixed with aqueous potassium iodide (a colourless solution) it will at first sink to the bottom as a green layer. If shaken, however, the reactants come into contact and the iodine is displaced by the chlorine, turning the aqueous layer brown. Some of this newly released iodine also dissolves back into the organic solvent layer, turning it from green to purple. Bromine gives an orange–brown colour in this solvent.

This method was used to view the displacement reactions between halogens and halides. Copy and complete the table below to show the final colour seen in the tetrachloroethene layer.

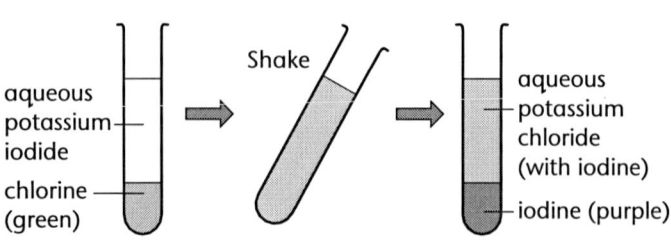

	Cl^- (aq)	Br^- (aq)	I^- (aq)
Chlorine in solvent			purple
Bromine in solvent			
Iodine in solvent			

4 Read the section in the text book which describes the extraction of bromine from sea water.

  a Bromine is obtained from the bromide ions in sea water by **i** oxidation, **ii** reduction and then **iii** oxidation again. Describe the chemistry of these three stages.
  b Describe the physical processes that accompany steps **ii** to **iii**, that allow the bromine to be separated from sea water, concentrated and finally purified.

© Farrow R, Gibbens D, Stirrup M and Vowles R, 2000. AS Chemistry for AQA – Resource Pack, Heinemann

Halides are reducing agents

① Halides are reducing agents. That is, they are electron donors. For example, iodide ions will reduce copper(II) ions to copper(I) ions:

Half-equations: $I^- \rightarrow \frac{1}{2}I_2 + e^-$ iodide ion loses electron (OIL – oxidation)

$I^- + Cu^{2+} + e^- \rightarrow CuI$ copper(II) ion accepts electron (RIG – reduction)

Added together: $Cu^{2+} + e^- + I^- \rightarrow CuI + \frac{1}{2}I_2 + e^-$

Cancelling the e^-: $Cu^{2+} + 2I^- \rightarrow CuI + \frac{1}{2}I_2$

Use the method above to show how reduction and oxidation occur in the following halogen reactions:

   a Magnesium bromide reduces chlorine gas to chloride ions

   b Silver chloride forms metallic silver when exposed to light

   c Manganese(IV) ions are reduced to manganese(II) ions when concentrated hydrochloric acid is added to manganese(IV) oxide. (Look at the Mn ions only.)

② **a** What is the trend in reducing ability of the halides down Group VII?

   b Explain this trend in terms of ionic radius and the ease with which electrons are lost.

③ Sodium chloride, sodium iodide and sodium bromide are all white crystalline salts. Three watch glasses with these salts lost their labels and were relabelled A, B and C. When treated with concentrated sulphuric acid:

A: gave off brown fumes and a choking odour was detected
B: gave off misty fumes
C: gave off purple fumes and a smell of rotten eggs was detected.

Which was which? Explain your answer.

Reaction with concentrated sulphuric acid

④ Look at these equations for the most extreme reaction between concentrated sulphuric acid and the halide ions:

A: $KCl + H_2SO_4 \rightarrow KHSO_4 + HCl$ (the same reaction occurs with KBr and KI)

B: $2HBr + H_2SO_4 \rightarrow Br_2 + SO_2 + 2H_2O$ (the same reaction occurs with HI but not HCl)

C: $8HI + H_2SO_4 \rightarrow 4I_2 + H_2S + 4H_2O$ (occurs with HI only)

Copy and complete the table below to show the oxidation numbers for sulphur and note whether oxidation is taking place: From this trend, deduce the relative reducing power of Cl^-, Br^- and I^-. Explain your answer.

	Oxidation number of the sulphur atoms		
Reaction	Reactants	Products	Is the sulphur reduced?
A			
B			
C			

© Farrow R, Gibbens D, Stirrup M and Vowles R, 2000. AS Chemistry for AQA – Resource Pack, Heinemann

Which halide?

1 a Copy and complete the table below, showing the reaction between halide ions and silver nitrate solution followed by the addition of ammonia solution.

Halide ions (aq)	With AgNO$_3$(aq)	Plus dilute NH$_3$(aq)	Plus concentrated NH$_3$(aq)	With Pb(NO$_3$)$_2$(aq) (see part b)
F$^-$	No reaction			
Cl$^-$	White precipitate			
Br$^-$			Precipitate dissolves	
I$^-$				

b Lead halides are all insoluble in water. Lead fluoride and lead chloride are white, while lead bromide and lead iodide are yellow. Complete the fifth column of your table to show what would happen if solutions of the different ions were added to lead nitrate solution.

c A lithium halide gave no reaction with silver nitrate solution but a white precipitate with lead nitrate solution. What is the salt?

d A magnesium halide gave a creamy precipitate with silver nitrate solution but a yellow precipitate with lead nitrate solution. What is the salt? What test could you do on the precipitate with silver nitrate to confirm this?

e The white precipitate given with potassium chloride and silver nitrate turns purple–grey when exposed to sunlight. Explain what is happening.

f A metal halide gave a yellow precipitate with silver nitrate solution and a yellow precipitate with lead nitrate solution. It also gave a yellow colour when held in a Bunsen burner flame. What is the salt?

Silver halides in photography

2 Read this passage about black and white photography and answer the questions below. Photographic film is coated with a layer of silver bromide in gelatin. When light falls on this, some of the silver bromide breaks down to silver metal. The film is then put into 'developer' solution (a mild reducing agent) which makes more metallic silver form around the 'light-produced' silver atoms. This makes the picture darker so that you can see it. Any unchanged silver bromide is removed using 'fixer' solution, which carries the remaining silver ions away.

a Explain the chemistry of the first two stages of this process.

b What would happen if the 'fixer' solution was not used?

c Why must the film then be printed onto (similar) photographic paper before the image is viewed?

Halogens in water

1. Bromine dissolves in water to give a brownish-coloured equilibrium mixture:
   $Br_2 + H_2O \rightleftharpoons 2H^+ + Br^- + BrO^-$
   a. What are the oxidation numbers of bromine in Br_2, Br^- and BrO^-?
   b. Explain in redox terms what happens to the bromine when it forms bromide ions.
   c. Explain in redox terms what happens to the bromine when it forms bromate(I) ions.
   d. Use this example to explain what the term **disproportionation** means.

2. Bromine water is an equilibrium mixture (see question 1). Use Le Chatelier's Principle to explain:
   a. why pH paper dipped in bromine water first turns red and then white.
   b. why adding alkali to the mixture makes the brown colour disappear.
   c. why adding acid to the mixture makes the brown colour stronger.

3. Bleaching powder is made by turning *slaked lime* (calcium hydroxide, $Ca(OH)_2$) in an atmosphere of chlorine.
   a. Use your understanding of the chemical reaction between chlorine and sodium hydroxide to predict the two main products of this reaction.
   b. Write a balanced chemical equation for this.
   c. Why should bleaching powder never be used with acidic cleaners?

Manufacture and use of chlorine

4. The electrolysis of sea water produces chlorine at the anode, hydrogen at the cathode and leaves sodium hydroxide behind in solution.
   a. At the anode: $2Cl^- \rightarrow Cl_2 + 2e^-$
      Is this oxidation or reduction? Explain your answer.
   b. Hydrochloric acid is made by burning hydrogen in chlorine and dissolving the product in water. Write an equation for this reaction, showing where oxidation and reduction are occurring.
   c. Describe how bleach may be made from sea water.
   d. Bleach contains sodium chlorate(I). This breaks down to sodium chloride and sodium chlorate(V) on heating.
      $3NaClO \rightarrow 2NaCl + NaClO_3$
      Use the oxidation numbers of the chlorine in this reaction to show that this is another example of disproportionation.

5. Chlorine and fluorine were once widely used to make organic chlorofluoro compounds (CFCs). What were they used for and why are they less commonly used today?

Module 2 — Foundation physical and inorganic chemistry — 9 Group VII, the halogens

Value for money

1

a. What is the 'active ingredient' in bleach? What does this liberate when it reacts?

b. The table below lists the stages used to calculate the amount of 'available chlorine' in bleach. Copy the table and complete the second column to explain why, in general terms, each step is taken.

1 A known volume of bleach is taken	
2 An excess of potassium iodide is added	
3 Sulphuric acid is added	
4 This mixture is titrated against sodium thiosulphate solution of know molarity until the yellow colour disappears	
5 This is repeated, this time adding starch solution just before the end point	

c. Why is the potassium iodide added *before* the acid? (Why would it be hazardous if the acid was put in first?)

d. Why is starch added?

2 Look at this table of a student's test results on some different bleaches. The molarity of the 'available chlorine' is in $mol\,dm^{-3}$ (or $mol\,litre^{-1}$).

Name	Price/£	Bottle volume /cm^3	Calculated molarity	Mol pound^{-1}
Superthick bleach	1.75	1000	0.15	
Supermarket best	2.12	2000	0.12	
Cheapo bleach (new stock)	1.29	1136	0.09	
Cheapo bleach (old stock)	1.29	1136	0.075	

Using a spreadsheet (or calculator) work out the mol pound^{-1} values for the final column, and so work out which bleach gives the best value for money.

3

a. In a test on bleach, a $25\,cm^3$ sample was diluted up to $500\,cm^3$ with distilled water. A $25\,cm^3$ sample of this diluted bleach was then transferred to a flask, and potassium iodide and acid were added. This was then titrated against $0.01\,M$ thiosulphate solution, using starch indicator. This was repeated until concordant results were obtained, and the average of these results was $23.55\,cm^3$. Calculate the molarity of the bleach by copying and completing the steps below.

Number of moles of thiosulphate = volume × molarity/1000 = ...

But 2 moles of thiosulphate are needed for every 1 mole of iodine (hence chlorine) so:

Number of moles of chlorine in the $25\,cm^3$ dilute bleach sample = ...

Number of moles in $500\,cm^3$ of dilute = $25\,cm^3$ of original bleach = ...

Therefore the concentration of chlorine in the original bleach ($mol\,dm^{-3}$) = ...

b. Which of the bleaches from question 2 is this most likely to be?

© Farrow R, Gibbens D, Stirrup M and Vowles R, 2000. AS Chemistry for AQA – Resource Pack, Heinemann

Principles of extraction

① What are:
  a the two most abundant elements in the Earth's crust
  b the two most abundant metals in the Earth's crust
  c the important physical properties of metals?

② Explain why copper and nickel are commonly used metals even though they are both relatively rare in the Earth's crust.

③ a State the chemical reaction always involved in the extraction of a metal from its ore.
  b Give three factors which must be taken into account when choosing the method of extraction.
  c Carbon is a common reagent used to extract a metal from its oxide.
    i Give a reason for its common use.
    ii Its use can lead to the production of carbon monoxide and carbon dioxide. State one problem associated with **each** of these by-products.
  d Hydrogen is sometimes used to extract a metal.
    i Suggest why hydrogen is used to extract tungsten but not iron.
    ii Suggest **two** advantages and **two** disadvantages of using hydrogen other than its cost.
  e Some metals are obtained by an electrolytic method.
    i Why might this be necessary?
    ii State one essential condition for such a method.
    iii Give one general disadvantage of electrolysis.

④ a Name a common ore from which iron is extracted.
  b State an essential condition required for the process.
  c Give the **three** other raw materials.
  d Write an equation to show:
    i the production of the gaseous reducing agent
    ii the reaction of the iron ore with this reducing agent
    iii the reduction of the iron ore by one of the raw materials.
  e The use of one of the raw materials has not yet been referred to.
    i Write an equation to show its decomposition.
    ii Give an impurity which can be removed by reaction with the product of the decomposition in part **i**.
    iii Write an equation for this reaction.
    iv Give a simple chemical reason why this can happen.
    v Give a use for the product of the reaction in part **iii**.
  f The iron obtained is not pure.
    i State **three** impurities present in the iron.
    ii Describe briefly how these impurities are removed.
  g The purified iron is converted into steel. By reference to the properties and uses of the impure iron and steel explain why such a process is necessary.

Module 2 — Foundation physical and inorganic chemistry — 10 Extraction of metals

Aluminium and titanium

①   **a**   Theoretically carbon could be used to obtain titanium from its oxide TiO_2. Explain why, in practice, it cannot. Write an equation to illustrate your reason.

      b   Similarly, tungsten cannot be obtained from its oxide using carbon. Instead, hydrogen is used. Write an equation for the reaction of tungsten(VI) oxide with hydrogen.

②   **a**   Name an ore from which aluminium is obtained.
      b   Suggest why this ore is purified prior to the electrolytic extraction process.
      c   State an essential condition required for the electrolysis.
      d   Say why cryolite (Na_3AlF_6) is added to the aluminium ore.
      e   What are the electrodes made of ?
      f   Write an equation for the reaction at:
          i   the cathode
          ii   the anode.
      g   By reference to the reaction occurring at the anode give a disadvantage of the process.
      h   Suggest the most important reason for The British Aluminium Company setting up plants in the Highlands of Scotland.
      i   State **three** properties of aluminium and for each property give a use which illustrates that property.
      j   Iron and aluminium are both recycled. Give **three** reasons why this is desirable.

③   **a**   Name an ore of titanium in which it occurs as TiO_2.
      b TiO_2 is not directly reduced to titanium. Show, by means of an equation and stating the essential conditions, how the next step involves converting it into another titanium compound.
      c   The final step is the reduction of this second compound.
          i   State **two** reducing agents which could be used.
          ii   Give the essential conditions.
          iii   Write an equation for the reaction.
      d   The extraction process has high costs despite titanium having a high natural abundance. State **four** factors which contribute significantly to the high cost of the process.
      e   Titanium is a desirable engineering metal.
          i   State **two** properties which make it a very useful material in the aircraft industry.
          ii   State one other property that allows it to be used in artificial hip joints.
          iii   Whereas impurities in iron can be tolerated, titanium must be pure. Explain why.
      f   "If we could find a suitable solvent in which TiO_2 would dissolve below 1000 °C, the iron and steel industry would be wiped out overnight." Discuss the validity of this statement. (What property must the 'suitable solvent' possess?)

④   **a**   What do you understand by the terms *'continuous process'* and *'batch process'*? Include in your answer the implications for the cost of a process.
      b   Into which category do the extractions of iron, aluminium and titanium fall?

Methods of extraction

1. Read the following passage about the extraction of sodium and then answer the questions that follow.

   Sodium is extracted from sodium chloride by an electrolytic method. The cell has a graphite anode and a steel cathode. Sodium chloride, to which has been added some calcium chloride, is heated to melting point (about 600°C); the molten state is then maintained by the heat generated by the passage of a very large current (20 000 amps). Sodium and chlorine are produced at the electrodes and kept apart by a system of hoods.

   a Why is sodium extracted by electrolysis?
   b Suggest why calcium chloride is added.
   c Why can the current not be passed initially, heating the sodium chloride and calcium chloride mixture up to 600°C?
   d At which electrode is:
      i sodium
      ii chlorine
      produced?
   e Construct the electrode equations.
   f Suggest why the sodium and chlorine must be kept apart.
   g Sodium chloride solution can also be electrolysed. Although chlorine is produced, sodium is not; instead another very important chemical is produced. Suggest a reason why this is so. What do you think the other chemical is?
   h The centre of the so-called chlor-alkali industry based on sodium chloride as a raw material developed in North Cheshire. Suggest the major reason for this.

2. The chief ore of zinc is zinc blende which is mainly the sulphide ZnS. It is converted into zinc oxide by roasting it in air; the sulphur is converted into sulphur dioxide. The zinc oxide is then heated with coke in a blast furnace at 1000°C. At this temperature some of the coke burns to give carbon monoxide.

   a Write an equation for the conversion of zinc sulphide into zinc oxide.
   b Suggest one problem arising from this reaction.
   c Write an equation for the reaction of zinc oxide with coke.
   d Suggest one problem arising from this reaction.
   e Write an equation for the reaction of zinc oxide with carbon monoxide.
   f The melting point of zinc is 419°C and its boiling point is 907°C. Suggest the method used to remove the zinc from the furnace.
   g During the 1950s there was a world surplus of zinc and it became almost uneconomic to extract it. With reference to the roasting process, suggest why the extraction was nevertheless continued. *(Hint: think about one of the by-products.)*

Module 2 | Foundation physical and inorganic chemistry | 10 Extraction of metals

Preparing reports

The following four questions can be tackled in small groups. They involve you in writing a short report and giving a presentation to the rest of the class.

1. You are commissioned to carry out a feasibility study concerning the search for and exploitation of a metal ore.

   Prepare a report discussing the essential points to be considered before a final decision on whether or not to go ahead would be made.

   Your report should consider:

   a an overview of the whole process, e.g. search, obtaining the ore, extraction of the metal, purification, uses of the metal
   b the costs for each stage.

2. Prepare a report on the factors you would consider in order to select the best possible site for an aluminium extraction plant in the United Kingdom. Amongst other things you should include considerations of cost (lots to think about there!), effects on the environment, availability of labour and the effects on the local residents.
   The following points may be of use:

   a Aluminium is the most abundant element in the Earth's crust (8%). Its main ore contains about 60% Al_2O_3 – the rest being Fe_2O_3 and SiO_2 – and is found mainly in Australia, Jamaica and West Africa.
   b The melting point of Al_2O_3 is 2072 °C.
   c It requires about 14 kW hours to produce 1 kg of aluminium. (That's a lot of energy!)

3. Prepare a report on the recycling of aluminium. Present it in two parts.

   Part One – Is it necessary?

   Consider the following points:

   a It costs less to re-use aluminium than to extract it. Melting down a can takes only about 5% of the energy needed to make a new one.
   b What happens to the aluminium if it is not recycled?

   Part Two – How is it done?

   Consider the following points:

   a How would you collect scrap aluminium? Could you keep it apart from other scrap? Could you organise collection points at your school or college?
   b How would the aluminium be transported to a recycling plant? Where might that be?

4. You are asked to present a report to a potential buyer of a local scrap-yard. He wants to know how easy (cost effective) it would be to recycle metals. Cars are common sources of scrap (mainly iron and steel), but they may contain other metals as well as plastics etc.

   Consider the following points:

   a Where does the scrap metal come from? Do you know what the metals are? What about alloys?
   b Which metal would be the most profitable? Who will buy it?
   c What are the day-to-day running costs?

© Farrow R, Gibbens D, Stirrup M and Vowles R, 2000. AS Chemistry for AQA – Resource Pack, Heinemann

Formulae of organic compounds

① Match each one of the following to one of the sentences below:
   1 empirical formula,   2 molecular formula,   3 structural formula,
   4 homologous series,   5 functional group,   6 structural isomers.

   A Shows the number of atoms of each element present in one molecule of a chemical compound.
   B A group of atoms that determines a compound's chemical characteristics.
   C A formula that shows how atoms are bonded together in a molecule of a chemical compound.
   D Molecules having identical molecular formulas, but their structural formulas are different.
   E Shows the simplest ratio of atoms of different elements present in a chemical compound.
   F A family of chemical compounds with the same general molecular formula.

   > Revise your work on Unit 4 and read about geometrical isomerism (Unit 13) before attempting questions 2 to 4.

Empirical and molecular formulae and isomers

② Combustion analysis was used to determine the empirical formula of an organic compound, W. The relative molecular mass of W was determined by a mass spectrometer. Use the data below to calculate the empirical and molecular formula of W.

   %C = 52.174%, %H = 13.043, %O = 34.783. Molecular ion peak at m/z = 46

③ Compound X has the molecular formula C_3H_6O. Draw:
   a two structural isomers of the compound that have one carbonyl group (>C=O) present in the molecule. Name each isomer.
   b a structural isomer (non-cyclic) that contains two functional groups one of which is an ether.

   > Ethers contain C—O—C

   c How many isomers can be drawn containing a ring of atoms? Draw the structures.
   d Three structural isomers are enols (an enol contains an —OH and a >C=C< functional group). One of them shows geometrical isomerism. Draw geometrical isomers that are enols. Name the enols which do not show geometrical isomerism.

   > Hint – There are two different groups on each end of a >C=C< for cis-trans isomerism.

Structural isomerism

1 An analysis was carried out on a hydrocarbon.

a Copy this table and fill in the missing information.

Element	g/100g	Moles	Mole ratio
Carbon	88.235	?	?
Hydrogen	?	11.765	?
$M_r = 68$	$(C_?H_?) \times 1 = 68$	Molecular formula = ?	

b What is the greatest number of rings that could be present in one of the structural isomers of this compound?

c Draw a cyclic isomer that has two double bonds.

d Draw cis- and trans- isomers of the compound that has two double bonds.

2 Three organic compounds were burnt completely in oxygen. One mole of each compound burnt to produce 5 moles of water and 5 moles of carbon dioxide. The compounds contain hydrogen and carbon only.

a What does this tell you about the three compounds?

b What is the molecular formula of each compound?

c The Baeyer test can be used to detect an unsaturated hydrocarbon. One of the three compounds did not give a positive result with the test. Of the two that did react, one compound shows cis-trans isomerism. Suggest one structural formula for each compound, including all geometrical isomers.

Homologous series and structural isomers

3 The table below shows the molecular formulae and boiling points of a family of organosulphur compounds.

a Write a general formula for this homologous series.

b Plot a graph of boiling point against number of carbon atoms present in the molecule. Can you explain the general trend in the graph?

Compound	Boiling point/°C
C_2H_6S	35
C_3H_8S	68
$C_4H_{10}S$	98.5
$C_5H_{12}S$	?
$C_6H_{14}S$	151
$C_7H_{16}S$	177
$C_8H_{18}S$	199

c Use your graph to estimate the boiling point of $C_5H_{12}S$.

d In these organosulphur compounds sulphur is able to form two single covalent bonds to two separate atoms. For the molecular formula $C_4H_{10}S$ draw two (position) structural isomers.

e Draw two (functional group) structural isomers for $C_4H_{10}S$.

© Farrow R, Gibbens D, Stirrup M and Vowles R, 2000. AS Chemistry for AQA – Resource Pack, Heinemann

Empirical, molecular and structural formulae

① A group of fluorohydrocarbons can be represented by the general formula $C_nH_{2n}F_2$. One member of the family has a relative molecular mass of 94. What is the molecular and empirical formula of the compound?

Structural formulae from names

② Draw a structural formula for each of the following compounds:
   a 1-aminopropane
   b 3-ethyl-2,2-dimethylpentane
   c 2-methylpropan-2-ol
   d 2-chloro-4-hydroxypentanoic acid
   e cyclopropylmethanol
   f propane-1,2,3-triol
   g ethylcyclohexane
   h 1-chloro-2-methylpropane
   i 3-methylpent-1-ene
   j 3-methylbutanone
   k cyclopentene
   l 1,6-diiodohexane
   m 2,3-dimethylbut-1-ene
   n pentanedinitrile

③ A student incorrectly named some organic compounds. These are listed below. Draw a structural formula for each compound and label it with the correct IUPAC name.
   a 1,1-dimethylbutane
   b 3-chlorobutane
   c 2-ethylpropane
   d pentan-4-one

④ Which structures are identical in **a** and **b**?

a

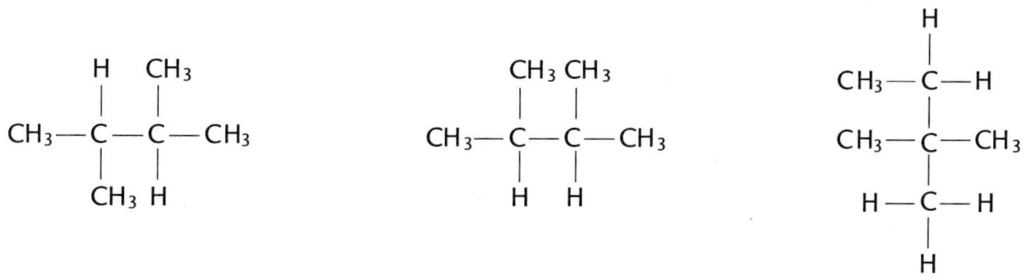

```
      H   CH3                CH3  CH3                      H
      |   |                   |    |                       |
CH3—C—C—CH3            CH3—C—C—CH3              CH3—C—H
      |   |                   |    |                       |
     CH3  H                   H    H                 CH3—C—CH3
                                                           |
                                                     H—C—H
                                                           |
                                                           H
```

b

```
  CH3  H                CH3  OH              OH  CH3
   |   |                 |    |               |    |
   C = C                 C = C                C = C
   |   |                 |    |               |    |
  CH3  OH                H    H               H   CH3
```

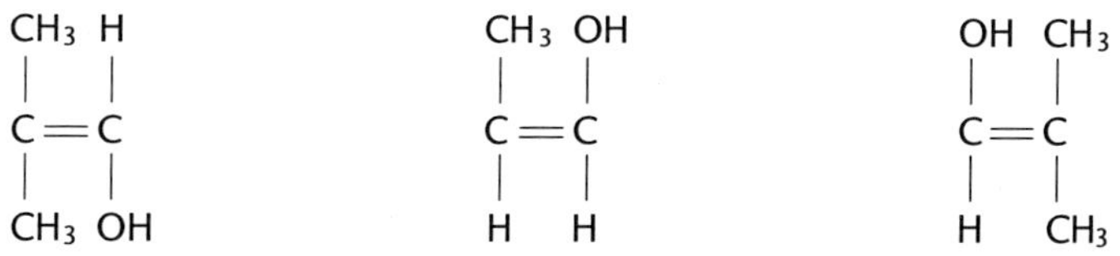

⑤ Draw all the structural isomers for the monosubstituted haloalkane $C_5H_{11}Br$. Name each isomer.

Naming more organic compounds

1. Name the following compounds.

a 3-chlorobutan-2-ol? — structure: H-C-C-C-C-H with HO, H, Cl, H substituents (butan-2-ol with Cl on C3): CH₃-CH(OH)-CH(Cl)-CH₃

b structure with C=C, CH₃ and two C-H branches

c H-C-C-C-C-C=O with Br on C3 (pentanal derivative)

d H-C-C-C-C-H with CH₃ branch

e H-C-C-C-C-C≡N with OH

f H-C-C-C-H with I substituent

g H-C-C-C-C-C-H with CH₃ and Cl branches

h H-C-C=O with Br, OH, Cl substituents

i H-C-C-C-C-C≡N with two CH₃ branches

j C=C-C=O with H substituents

(Structural diagrams shown for naming practice)

© Farrow R, Gibbens D, Stirrup M and Vowles R, 2000. AS Chemistry for AQA – Resource Pack, Heinemann

Fractional distillation and cracking

Fractional distillation

The components of petroleum can be separated into a number of fractions by fractional distillation. A simplified diagram of a fractionating column is shown below.

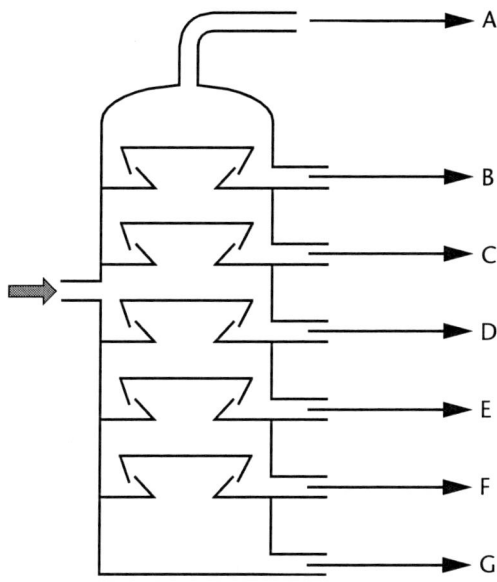

① What is fed into the column?

② What property of the substances allows them to be separated?

③ Name four of the fractions labeled B to F and give a use for each.

④ How does the temperature change as you go up the column?

⑤ Explain why some of the liquids inside the column boil even though the column is not heated.

⑥ There are some molecules with twelve carbon atoms in fractions C and D. How is this possible?

Cracking

There is a large demand for fuels. The lighter fractions make the best fuels so there is a greater demand for these fractions. The cracking process is used to convert some of the heavier fractions, present in petroleum in relative excess, to lighter ones to make petrol.

⑦ State the conditions and describe what happens in the cracking process.

⑧ The bond energies for the C—C and C—H bonds are $347\,\text{kJ}\,\text{mol}^{-1}$ and $413\,\text{kJ}\,\text{mol}^{-1}$ respectively. Which type of bond would you expect to break more readily when a mixture of hydrocarbons is heated strongly? Explain your answer.

© Farrow R, Gibbens D, Stirrup M and Vowles R, 2000. AS Chemistry for AQA – Resource Pack, Heinemann

Industrial applications

① The following compounds are present in petroleum.
Copy out these structures and name each one.

a
```
        CH₃
        |
CH₃—C—CH₂—CH₂—CH₂—CH₃
        |
        CH₃
```

b
```
           CH₃ CH₃
            |   |
CH₃—CH₂—C—CH—CH₃
            |
           CH₃
```

c
```
       CH₃ CH₃
        |   |
CH₃—C—C—CH
        |   |
       CH₃ CH₃
```

d
```
                CH₃
                 |
CH₃—CH₂—CH₂—CH—CH₂—CH₂—CH₃
```

② If a mixture of these four compounds was fractionally distilled give the order in which they would distil off and explain your answer.

③ Name the type of isomerism they show.

④ Given that the bond energy for the C—C bond is 347 kJ mol⁻¹ and that for the C—H bond is 413 kJ mol⁻¹ what would you expect to happen when a mixture of hydrocarbons is heated strongly?

⑤ Which industrial process relies on this?

⑥ Write an equation to show what could happen when a hydrocarbon with 16 carbon atoms undergoes homolytic fission.

⑦ As well as producing smaller alkanes, members of another homologous series are formed. Define the term homologous series and name the series to which the other hydrocarbons belong.

⑧ A straight-chained alkane with 16 carbon atoms forms three straight-chained hydrocarbons each with a different number of carbon atoms. Write an equation for one possible reaction and explain why it is not possible for all of the products to be alkanes.

Combustion of fuels

> Octane, C_8H_{18}, is a component of petrol.

① Write an equation showing the complete combustion of octane.

② Why is it that when octane burns in a car's engine combustion is often incomplete?

③ Incomplete combustion leads to the formation of a harmful product. Write an equation showing incomplete combustion of octane.

④ Identify the harmful product and say why it is harmful.

⑤ Name another harmful product that can form when petrol burns in the engine of a car and say why it is harmful.

⑥ How are the amounts of these harmful substances reduced before the exhaust fumes are allowed to escape into the atmosphere?

⑦ Write an equation or equations to show the conversion of these harmful substances into less harmful ones.

⑧ When combustion is incomplete some of the alkanes in the petrol remain unburned. If these substances are released into the atmosphere, what effect will they have?

⑨ Until recently, tetraethyl lead was added to petrol to prevent 'knocking'. Why is it not possible to use leaded petrol in cars produced in the last few years?

⑩ Why is it considered to be desirable to limit the use of cars? Suggest three ways in which the number of car journeys could be reduced.

> Sulphur is a contaminant of petroleum. Almost all of the sulphur is removed from petrol during the refining process but it is not removed from the oil used in oil-fired power stations.

⑪ What is most of the removed sulphur used to make?

⑫ Write an equation to show what would happen to any sulphur left in the petrol or oil when the fuel is burned. Explain the environmental effects of allowing the substances formed when sulphur burns to escape into the atmosphere.

Petroleum

Cracking is an important industrial process. It is used to convert some of the heavier fractions, obtained from fractional distillation of petroleum, into lighter ones, for which there is more demand. There are two different types of cracking: thermal cracking and catalytic cracking.

Thermal cracking

1. What are the conditions for thermal cracking?

2. What type of bond fission occurs?

3. Name the type of chemical species produced by this type of bond fission and say what feature all members of this species have.

4. Why do C—C bonds tend to break more readily than C—H bonds during this process?

Catalytic cracking

5. What are the conditions for catalytic cracking?

6. Which organic species are formed as intermediates during this process?

7. If you wished to produce alkenes for use in organic synthesis which of the two cracking processes would you use?

8. What is the other type of cracking used to produce?

> If methane and chlorine are mixed, in the dark, at room temperature, there is no reaction. If the mixture is exposed to bright sunlight it explodes.

9. Write equations to show the stages of the reaction between methane and chlorine and label each of the stages.

10. If the mixture contained a large excess of methane what would be the major product?

11. Why is this reaction not considered to be a good way of making chloromethane?

Unsaturated hydrocarbons

① Which of the following straight-chain/non-cyclic molecules are alkanes and which are alkenes. Explain your answer.
   a C_6H_{14} b C_3H_6 c C_3H_8 d C_5H_{12} e C_4H_8

② Name and draw the alkenes you have identified in question 1.

③ Write the molecular formulae for:
   a ethene   b pentene.

④ The formula of hexene is C_6H_{12}.
   a Hexene has three *positional isomers*. Explain what is meant by this.
   b Name the following positional isomers of hexene. (Drawn as 90° structural formulae.)

   i
   ```
        6  5  4  3  2  1
        H  H     H  H
        |  |     |  |
    H—C—C—C=C—C—C—H
        |  |  |  |  |  |
        H  H  H  H  H  H
   ```

   ii
   ```
        H  H  H        H
        |  |  |        |
    H—C—C—C—C=C—C—H
        |  |  |  |  |  |
        H  H  H  H  H  H
   ```

   c Name these isomers (careful).

   i
   ```
        H  H  H  H
        |  |  |  |
    H—C—C—C—C—C=C—H
        |  |  |  |  |  |
        H  H  H  H  H  H
   ```

   ii
   ```
              H  H  H  H
              |  |  |  |
    H—C=C—C—C—C—C—H
        |     |  |  |  |  |
        H     H  H  H  H  H
   ```

Geometrical isomers

⑤ a What are *geometrical isomers*?
   b What is different about a double bond (compared to a single C—C bond) that leads to the formation of geometrical isomers in some alkenes?
   c Name these two geometrical isomers of hexene. (Drawn to show the 120° double bond.)

   i
   ```
    H₃CH₂    CH₂CH₃
         \  /
          C=C
         /  \
        H    H
   ```

   ii
   ```
    H₃CCH₂      H
         \    /
          C=C
         /    \
        H      CH₂CH₃
   ```

⑥ a Name this isomer of hexene.

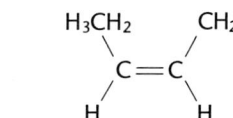

   b Write a balanced equation for its complete combustion in air.

⑦ Which of the following straight chain (non-cyclic) hydrocarbons will decolourise bromine water? Explain your answer.
   a ethane   b butene   c C_5H_{12} d C_7H_{14}

⑧ A straight-chain (non-cyclic) hydrocarbon with a 6-carbon atom chain decolourised bromine water. What is it called and what is its formula?

© Farrow R, Gibbens D, Stirrup M and Vowles R, 2000. AS Chemistry for AQA – Resource Pack, Heinemann

Breaking the double bond

Module 3 — **Foundation organic chemistry** — **13 Alkenes**

① Write the simple structural equation for the reaction of ethene with hydrogen chloride. What is this type of reaction called?

② Read this passage which explains the mechanisms involved in this type of reaction (question 1). Use this passage alongside your own knowledge to write brief definitions for the words in bold.

The difference between the alkanes and the alkenes is the presence of a carbon—carbon **double bond** in the latter. This makes the alkenes more reactive, as one part of the double bond is easier to break than a single bond, and this bond can 'snap open', giving rise to an **addition reaction** with another particle such as a chlorine molecule. Also, as the C=C bond is rich in electrons, it attracts electron deficient species called **electrophiles**, which then attack the bond and cause a reaction.

③ a What do the terms **dipole** and **carbocation** mean?
  b Copy the structural diagrams for the two-stage reaction of ethene with hydrogen bromide and annotate them to explain how the reaction takes place.

④ a Write a simple equation for the addition reaction between propene ($CH_3CH=CH_2$) and sulphuric acid. (Consider the major product only.)
  b What is the electrophile in this reaction?
  c What is the carbocation that forms as an intermediate in this reaction?
  d What will form if the product of the reaction in part **a** is warmed with water?
  e What is the process that occurs in part **d** called?

Induced polarity

⑤ a Chlorine molecules are **non-polar**. Explain this term.
  b Explain how and why a **temporary dipole** forms when a chlorine molecule approaches a C=C double bond.
  c Use these ideas to explain how 1, 2-dichloroethane forms from ethene and chlorine. Draw structural diagrams to help you.

⑥ Name the following compounds and describe how you could make each of them from the appropriate alkene.

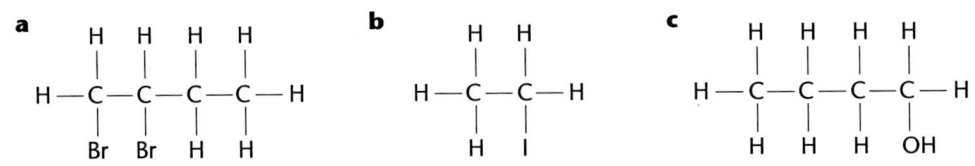

Major and minor products

1 a Name each of the alkenes shown below.

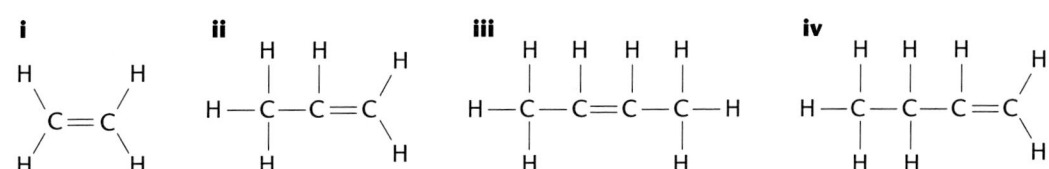

b Which of these alkenes will produce two different addition products with HBr?

2 a These three carbocations formed from different isomers of pentene. For each, state whether it is primary, secondary or tertiary.

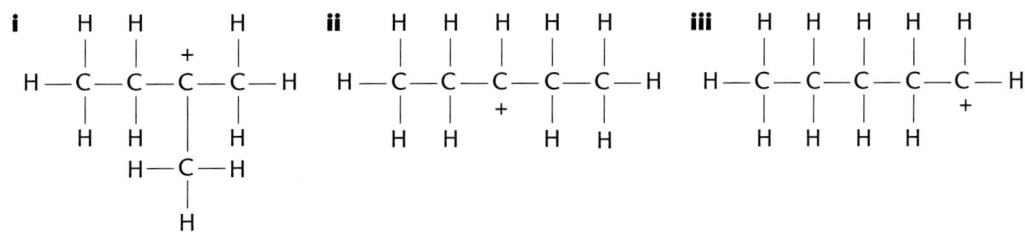

b Which of the three is the most stable, and why?

3 Pent-1-ene forms two different bromopentanes when reacted with HBr.

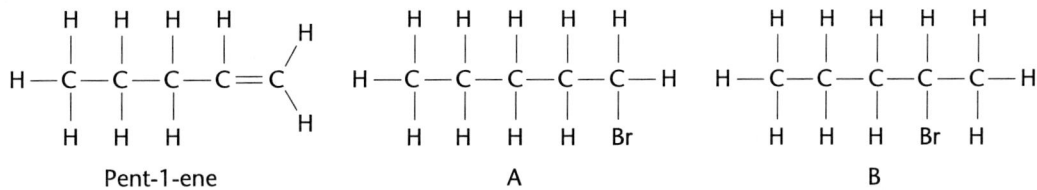

a Name the two products, A and B.
b Which will be the major product and which the minor product?
c Explain your answer to part **b**.

4 a Starting with pent-1-ene from question 3, describe how you could make pentan-1-ol and pentan-2-ol.
b Draw structural diagrams, showing the intermediary products.
c Which would be the major and which the minor product of the reaction?

5 Name each of the following substances. For each, state:
a which alkene they could be made from
b what reactant would be used
c whether they would be the sole product of the reaction or not
d if not the sole product, state whether they would be the major or minor product.

© Farrow R, Gibbens D, Stirrup M and Vowles R, 2000. AS Chemistry for AQA – Resource Pack, Heinemann

Hydrogenation

①
a Draw simple structural diagrams to show the reaction between ethene and hydrogen.
b What name is given to this type of reaction?
c What catalyst is used to make this reaction work?
d Hydrogen molecules are non-polar. How does this help to explain why ethene and hydrogen do not react without a catalyst, whereas ethene and hydrogen bromide do?

②
a Define the terms saturated and unsaturated as related to lipids (fats and oils).
b Animal lipids are solid at room temperature, whereas vegetable lipids tend to be liquid. What does this suggest about the level of saturation in their molecules?
c Describe how vegetable oils are 'hardened' to make margarine or chocolate.

Polymerisation

③
a Explain the terms **addition polymerisation, monomer** and **repeating unit**.
b Describe with the help of diagrams the reaction that occurs when ethene is made to polymerise to poly(ethene).
c Ethene is a gas: explain why poly(ethene) is a solid.

④ The strength of a polymer depends of the size of the intermolecular forces between the chain molecules. For any given polymer, this will also be affected by how closely the chains can pack together. Low density polythene chains have many branches, whereas high density polythene chains are relatively branch-free (as shown opposite). Which is likely to be the stronger? Explain your answer.

high density polythene

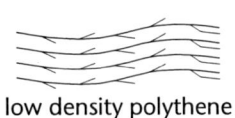
low density polythene

⑤ The diagram below shows part of a polymer chain used for making artificial fibres:

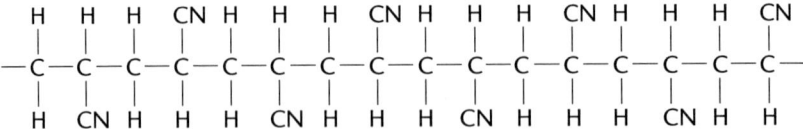

a Draw a structural diagram of the repeating unit of this polymer.
b Give the formula of the alkene that it is made from.
c Name the monomer (this type of nitrogen compound is called a nitrile) and so suggest a suitable name for the polymer.

⑥ Ethene will not normally polymerise spontaneously: the reaction is usually triggered by a free radical such as RO• where R is an akyl group and O• has a lone electron. Such a free radical will attack the C=C double bond, 'passing' the lone electron along the growing chain. Use the reaction sequence below to **explain** how ethene polymerisation is triggered, continues and is finally terminated.

Initiation: $RO^\bullet + C_2H_4 \rightarrow ROC_2H_4^\bullet$
Chain reaction: $ROC_2H_4^\bullet + C_2H_4 \rightarrow ROC_2H_4C_2H_4^\bullet$
  $ROC_2H_4C_2H_4^\bullet + C_2H_4 \rightarrow ROC_2H_4C_2H_4C_2H_4^\bullet$ etc. ...

Finally
Termination: $ROC_2H_4C_2H_4C_2H_4........C_2H_4C_2H_4C_4H_2C_2H_4C_2H_4C_2H_4C_2H_4C_2H_4^\bullet + {}^\bullet OR$
  $\rightarrow ROC_2H_4C_2H_4C_2H_4........C_2H_4C_2H_4C_2H_4C_2H_4C_2H_4C_2H_4C_2H_4C_2H_4OR$

© Farrow R, Gibbens D, Stirrup M and Vowles R, 2000. AS Chemistry for AQA – Resource Pack, Heinemann

Alcohols

① Alcohol (ethanol) was traditionally made by fermenting sugar with yeast in large vats and then distilling the resultant solution (it still is made this way for non-industrial uses!). Today most ethanol is produced by the catalytic hydration of ethene. Suggest what advantages the modern continuous production method has over the older batch method.

② Unlike many examples of catalysis, the mechanism by which phosphoric acid catalyses the hydration of ethene is relatively easy to understand. Draw structural diagrams to show the three-stage conversion of ethene to ethanol, under the following headings:

Electrophilic attack on the C=C bond by the acid
Carbocation formation
Reaction with water
Reformation of the acid

③ Describe how butan-2-ol may be made from both but-1-ene and but-2-ene (using concentrated sulphuric acid). In which case is it the *sole* product of the hydration of the alkyl hydrogensulphate?

Epoxyethane

④ a With the aid of this structural formula for epoxyethane, explain the term **cyclic ether**.

$$\begin{array}{c} \quad\; O \\ \quad / \; \backslash \\ H-C-C-H \\ \quad|\quad| \\ \quad H\; H \end{array}$$

b Why does this structure make epoxyethane so reactive?

⑤ Using structural diagrams, describe how antifreeze (ethane -1, 2-diol) may be made from ethene, noting the appropriate reaction conditions at each stage.

⑥ A polymer with the formula:

R—OCH$_2$CH$_2$—OCH$_2$CH$_2$—OCH$_2$CH$_2$—.....OCH$_2$CH$_2$—OCH$_2$CH$_2$—OCH$_2$CH$_2$—OH

is used to reduce the friction between the water and the nozzle in fire hoses, effectively doubling their range.

a What is the repeating unit in this polymer?
b What was the monomer reacted with to produce this polymer?

⑦ Ethan-1, 2-diol reacts with dicarboxylic acids to make condensation polymers such as terylene.

...HO—CH$_2$CH$_2$—OH + HOOC—R—COOH + HO—CH$_2$CH$_2$—OH + HOOC—R—COOH + HO—CH$_2$CH$_2$—OH...

→ ...O—CH$_2$CH$_2$—OOC—R—COO—CH$_2$CH$_2$—OOC—R—COO—CH$_2$CH$_2$—O...

(where **R** is a complex group)

a What small molecules are formed in addition to the main polymer?
b Suggest why this is called a condensation reaction.

© Farrow R, Gibbens D, Stirrup M and Vowles R, 2000. AS Chemistry for AQA – Resource Pack, Heinemann

Haloalkanes (1)

1 Write brief notes, including a definition if appropriate, on each of the following:
   a polar bond
   b nucleophile
   c nucleophilic substitution
   d elimination
   e base

2 Name each of the following haloalkanes:
   a $CH_3CHClCH_3$
   b $CH_3CH_2CHBrCH_2Br$
   c $CH_3CH(CH_3)CH_2CH_2I$
   d $CH_3CHClCH_2CH_2Br$
   e $CH_3CH_2CHBrCHBr_2$

3 Bromoethane can be made from ethene.
   a Write an equation for the reaction.
   b State the type of reaction involved.
   c Write the mechanism for the reaction.

4 Bromoethane can also be made from ethane.
   a Using your knowledge of the reactions of alkanes, suggest the other reactant needed and the essential condition required.
   b Write an equation for the reaction.
   c Suggest why this method may not be as good a method as that in question 2 for making bromoethane.

5 The following compounds can be made from haloalkanes. For each compound, name a haloalkane which could be used, write an equation and state the essential conditions.
   a propan-2-ol
   b 2-methylpropan-1-ol
   c 2-methylpropan-2-ol
   d propanenitrile
   e 2-methylpropanenitrile
   f ethylamine

6 For each of the following reactions, name the organic product and write an equation.
   a 2-bromobutane + dilute aqueous sodium hydroxide solution
   b 1-iodopropane + hot concentrated ethanolic potassium hydroxide solution
   c 1-iodobutane + excess ammonia
   d 1-bromopropane + potassium cyanide solution
   e 2-iodo-2-methylpropane + potassium cyanide solution
   f 1,4-dibromobutane + excess dilute sodium hydroxide solution
   g 1,4-dibromobutane + excess hot concentrated ethanolic potassium hydroxide solution
   h 1-bromocyclohexane + dilute sodium hydroxide solution
   i 1-bromocyclohexane + hot concentrated ethanolic potassium hydroxide solution

Haloalkanes (2)

1 **a** Draw the structural formula of bromoethane showing clearly the polar nature of the C—Br bond.

  b Choose from the following list which species could act as nucleophiles:
   H⁺   NH₃   NO₂⁺   Cl   AlCl₃   OH⁻   CH₃CH₂O⁻   CN⁻

2 **a** Draw the structures of the four isomers of C₄H₉Br. Name each compound.
  b Which are primary haloalkanes?
  c Which two of them could be converted into 2-methylpropene?
  d Which one can undergo elimination to give an alkene that shows geometrical isomerism?
  e 3-methylbutanoic acid can be made from a nitrile which can be made from one of the above isomers. Choose the correct isomer and write an equation for each reaction. Name the types of reaction occurring and show the mechanism for the formation of the nitrile.

3 The C—F, C—Cl, C—Br and C—I bond enthalpies are 484, 338, 276 and 238 kJ mol⁻¹ respectively. Use this data to decide which of fluoroethane and iodoethane reacts more quickly with dilute aqueous sodium hydroxide solution, and briefly explain your choice.

4 Compound **X**, C₂H₄Br₂, can be converted into compound **Y**, C₄H₄N₂, which when heated with hydrochloric acid gives butane-1, 4-dicarboxylic acid. Identify **X** and **Y**, write an equation for, and state the type of each reaction, and give the essential conditions for the conversion of **X** to **Y**.

5 **a** An iodoalkane **A**, with a molecular mass of 198, contains 30.3% carbon and 5.6% hydrogen by mass. It reacts with aqueous sodium hydroxide solution to give an alcohol which can be oxidised and with ethanolic potassium hydroxide solution to give **two** alkenes, neither of which show geometrical isomerism.
   i Identify **A**.
   ii Show the mechanisms of the two reactions referred to, stating the role of the hydroxide in each case.
   iii Write an equation for the reaction of **A** with ammonia and show the mechanism.
  b **A** has isomers **B, C, D, E, F, G** and **H**. Use the information below to identify them (**F, G** and **H** cannot be distinguished from each other).
   i **B** does not undergo elimination.
   ii **C** reacts with aqueous NaOH to give an alcohol which cannot be oxidised.
   iii **D** undergoes elimination to give only one alkene which shows geometrial isomerism.
   iv **E** undergoes elimination to give two alkenes only one of which shows geometrical isomerism.
   v **F, G** and **H** are all primary iodoalkanes which each undergo elimination to give only one alkene which does not show geometrical isomerism.

6 **Now practise writing mechanisms.** Write a mechanism for each of the reactions in questions 4 and 5. Draw your curly arrows carefully! Remember your lone pairs and charges!

Haloalkanes (3)

1. Compound **J**, $C_7H_{13}Br$, is a bromocyclohexane. It reacts with aqueous potassium hydroxide solution to give **K**, $C_7H_{14}O$, a tertiary alcohol. With hot ethanolic potassium hydroxide, two isomeric alkenes, C_7H_{12}, are produced. Identify **J** and **K** and the two alkenes. Write mechanisms to show the formation of the alcohol and the alkenes.

2. Compound **L**, C_2H_5BrO, reacts with potassium cyanide to give **M**, C_3H_5NO, which can be hydrolysed to 2-hydroxypropanoic (lactic) acid. Identify **L** and **M** and write an equation for each reaction.

3. The methoxide ion CH_3O^- and the ethanoate ion $CH_3-\underset{\underset{O}{\|}}{C}-O^-$ are both nucleophiles.

   Use your knowledge of nucleophilic substitution reactions to predict the product obtained when each of them reacts with bromoethane.

4. Compund **P**, $C_3H_5Br_3$, reacts with three moles of potassium cyanide to give **Q**, $C_6H_5N_3$. When heated with hydrochloric acid, **Q** reacts in three steps to give finally **R**, $C_6H_8O_6$. **P** reacts with aqueous potassium hydroxide to give **S**, $C_3H_8O_3$, and with ethanolic potassium hydroxide when boiled to give firstly **T**, $C_3H_4Br_2$, and with excess, **U**, C_3H_3Br. Compound **S** contains two primary and one secondary alcohol groups.

   a Give the structures of compounds **P**, **Q**, **R**, **S**, **T** and **U**.

   b Name compounds **P**, **S**, **T** and **U**.

   c Name the types of reaction in the conversions of **P** to **Q**, **Q** to **R**, **P** to **S** and **P** to **U**.

   d Write the mechanism for the conversion of **P** to **T**.

5. Suggest how the following syntheses can be carried out. Give essential reactants and conditions.

   a $CH_4 \rightarrow CH_3OH$ (two steps)

   b $CH_4 \rightarrow CH_3COOH$ (three steps)

   c $CH_3CH=CH_2 \rightarrow CH_3CH(CH_3)COOH$ (three steps)

   The following questions all involve haloalkanes but also incorporate other parts of the organic chemistry in Module 3.

6. Identify the unknown compounds in the following sequences:

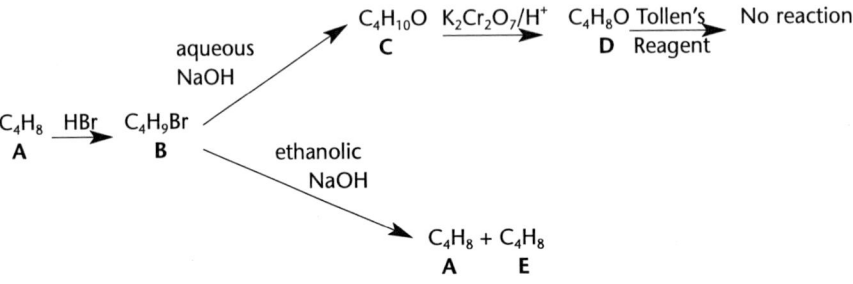

Compound A shows geometrical isomerism

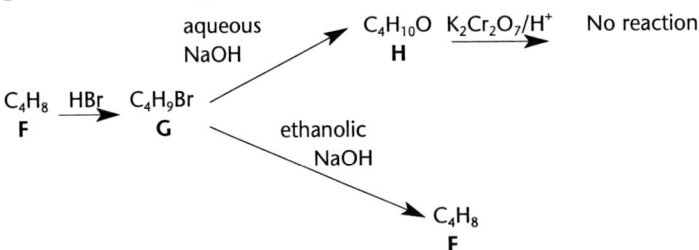

Alcohols and dehydration

Shown below are eight alcohols with the formula $C_5H_{11}OH$.

A $CH_3-CH_2-CH_2-CH_2-CH_2-OH$

B $CH_3-CH_2-CH_2-CH(OH)-CH_3$

C $CH_3-CH_2-CH(OH)-CH_2-CH_3$

D $CH_3-C(OH)(CH_3)-CH_2-CH_3$

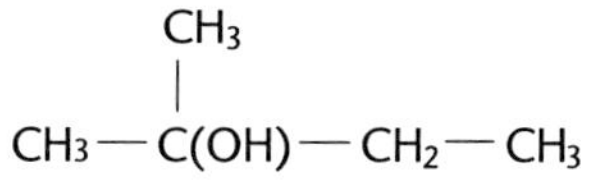

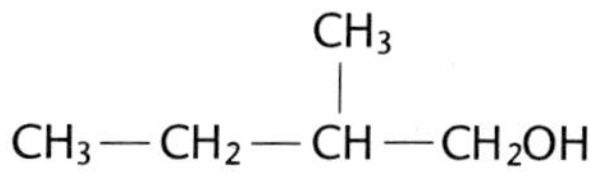

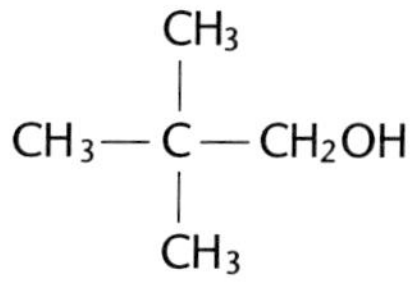

E $CH_3-CH_2-CH(CH_3)-CH_2OH$

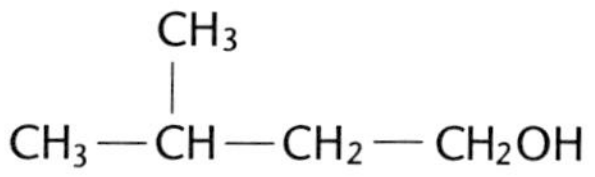

F $CH_3-C(CH_3)_2-CH_2OH$

G $CH_3-CH(CH_3)-CH_2-CH_2OH$

H $CH_3-CH(CH_3)-CH(OH)-CH_3$

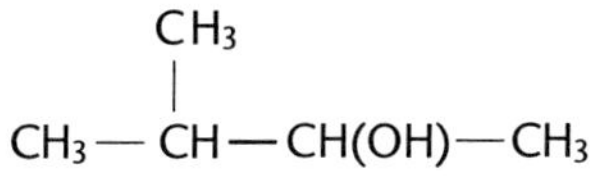

(1) Name alcohols **B**, **F** and **G**.

(2) If a mixture of all eight is fractionally distilled which one would distil off first and which could be left as a pure substance, in the flask, at the end?

(3) Which can be oxidized to give a carboxylic acid?

(4) Which does not cause warm acidified potassium dichromate(VI) to turn green?

(5) List the pairs that **can** be dehydrated to give the same alkene.

(6) Which can be dehydrated to give an alkene that can show geometrical isomerism?

(7) If you wanted to make **G** in the laboratory would it be better to start with 3-methylbut-1-ene or 1-bromo-3-methylbutane? Explain your answer with reference to the mechanisms of the reactions involved.

(8) One of the alcohols can be dehydrated to give *three* different products. Identify this alcohol by giving its structure or name and write out the structures of the three products.

© Farrow R, Gibbens D, Stirrup M and Vowles R, 2000. AS Chemistry for AQA – Resource Pack, Heinemann

Elimination and other organic reactions

This question brings together some of the organic chemistry from this and earlier sections.

① A compound, P, has the formula $C_4H_{10}O$. When P was dehydrated the compound Q, with the formula C_4H_8 was produced. When Q was reacted with an excess of bromine a third substance, R, was formed which reacted with an excess of warm, dilute, aqueous sodium hydroxide to give S, $C_4H_{10}O_2$. Partial oxidation of S gave T, $C_4H_6O_2$, which could be oxidized further to give W, which had the formula $C_4H_6O_3$. When P was oxidized, by adding it to an excess of warm acidified potassium dichromate(VI), X was produced. X had the formula $C_4H_8O_2$.

a Deduce the structures of the seven compounds and write equations for each of the reactions.

b Give the reagents and conditions for the dehydration of P.

c Write a mechanism for this reaction.

> Remember to use a curly arrow and to show bonds/lone pairs when necessary.

d The structures of four isomeric alcohols are shown.

A $CH_3-CH_2-CH_2-CH_2OH$

C
CH_3
|
$CH_3-CH-CH_2OH$

B $CH_3-CH(OH)-CH_2-CH_3$

D
CH_3
|
CH_3-C-OH
|
CH_3

Say which are positional isomers of each other.

e Which could be distinguished from the rest by addition of acidified potassium dichromate(VI)?

f Which could be oxidized to give a compound that reduces ammoniacal silver nitrate to give silver as a reduction product?

Ethanol production and reactions

Ethanol is an important chemical for both the chemical and the food and drinks industries. It can be manufactured by fermentation and by the direct hydration of ethene. Fermentation is a batch process whilst direct hydration is a continuous process.

1. Give an equation for each of the processes mentioned above and state the essential conditions.

2. Say what are the advantages of fermentation compared to direct hydration.

3. Which process is more likely to still be in use at the end of the century? Give reasons for your choice.

4. A simple way of increasing the rate of a reaction is to increase the temperature. Say why the temperatures of the fermentation and direct hydration processes are not increased above those you have given in your answer to question 1.

5. Ethanol can be oxidized by acidified potassium dichromate(VI) to give two different organic products. Write equations, using [O] to represent the oxidant, for each of these reactions.

6. Explain how the reaction could be carried out in two different ways in order to obtain each of these products.

7. Say how the purified products can be distinguished by giving a positive test for each.

> Propan-1-ol can be converted to propan-2-ol in two stages. The product of the first reaction is purified and then converted to propan-2-ol.

8. Write equations for the two reactions and give the conditions for each.

9. Give the mechanisms for each reaction.

10. By referring to the mechanism, explain why this method cannot be used to convert propan-2-ol to propan-1-ol.

Module 3 | Foundation organic chemistry | 15 Alcohols

Reactions of epoxyalkanes

Alcohols with two OH groups are called diols. Hydrating epoxyalkanes can form diols. Epoxypropane can be used to form two compounds that are biologically important plus a third that is of industrial significance. The structure of epoxypropane is:

$$CH_3-CH\underset{O}{-}CH_2$$

(1) Write an equation showing the formation of a diol from epoxypropane. You may want to check your notes and refer to Unit 13.

(2) **Complete** oxidation of the diol with acidified potassium dichromate(VI) gives one of the biologically important compounds.
 a Write an equation for this reaction, using [O] to represent the oxidizing agent.
 b Can you name this compound?

(3) **Partial** reduction of this compound can give the second biologically important compound.
 a Suggest a possible reducing agent and write an equation for this reaction, using [H] to represent the reductant.
 b Can you name this compound?

The second biologically important compound can be converted to a compound that is used in industry. This compound has a carboxyl group and an -ene group and it is a member of a series called either the alkene–monocarboxylic acids or the oleic acids.

(4) Suggest a reagent that could be used to bring this reaction about.

(5) Write an equation for the reaction and suggest the conditions that could be used.

(6) Give a mechanism for this reaction.

(7) How could you show that the final compound contains both an -ene bond and a carboxylic acid group?

(8) Write out a flowchart to show this reaction sequence. In your chart give the displayed structures of the compounds.

Classification and reactions

> An important difference between the various classes of alcohols is the way in which they react with oxidising agents.

① What are the three classes of alcohols?

② Members of which class can be oxidized to give products that are members of two different homologous series?

③ Which class of alcohols cannot be oxidized without some breakdown of their carbon skeletons?

④ Write out the structures of the straight-chained alcohols with the formula $C_5H_{12}O$.

⑤ Which, if any, of these alcohols can be dehydrated to give one product only?

⑥ Which, if any, can be dehydrated to give more than one product?

⑦ By referring to the mechanism of the reaction, explain your answers to questions 5 and 6.

> On analysis a monohydric alcohol was found to contain 60.00% carbon and 13.33% hydrogen.

⑧ Calculate the empirical and molecular formulae of this alcohol.

> When this alcohol was added to acidified potassium dichromate a compound was produced that did not react with either Fehling's solution or ammoniacal silver nitrate. When the compound was added to a solution of sodium hydrogencarbonate there was no reaction.

⑨ Write an equation for the reaction between the alcohol and the oxidant using [O] to represent the oxidant.

⑩ Give the structures of the alcohol and the product of the oxidation reaction.

Answers to worksheet questions

1 Atom size
1. 4×10^6
2. 16×10^{12}
3. 6×10^{23}
4. 6×10^{23}
5. a i 6+ ii 10+ iii 13+ iv 22+
   b Zero net charge on all atoms
6. Missing values from table are:

Sulphur	16	16	16
Calcium	40	20	20
Iron	26	26	30
Lead	82	125	82
Uranium	92	238	92

7. Three different elements: 18p22n, 19p21n and 20p20n, therefore very different properties (Ar, K, Ca)

2 Isotopes
1. Three isotopes of the same element (Ne): 10p10n, 10p11n, 10p12n, so same chemical properties
2. a Both oxygen isotopes but O-18 has two extra neutrons
   b $\times 1.125$
   c H has no protons, D has 1
   d $\times 2$
   e $\times 1.11$ therefore 11% denser ('heavier') than normal water
3. a Rn-220
   b H_2S – faster diffusion
   c U-235

3 Mass spectrometer
1. A The vaporisation chamber – vaporises the substance
   B The electron gun – fires high-speed electrons at the particles, creating cations
   C The ion accelerator – accelerates the ions in an electrostatic field
   E The deflection zone – ions deflected by strong electromagnetic field
   Y The detector – detects the ions
2. a Towards X
   b Towards Z
   c Towards X
   d No change
   e Towards Z
   f Towards Z
3. a 39:1   d 41
   b 39   e 7%
   c 93%

4 Working out masses
1. 63.62 (with appropriate working)
2.

	A	B	C	D	E	F	G	H
1			isotope 1		isotope 2		isotope 3	overall A
2	element	A	%	A	%	A	%	
3	Ga	69	60.4	71	39.6	0	0	69.79
4	Ag	107	51.35	109	48.65	0	0	107.97
5	Si	28	92.18	29	4.71	30	3.12	28.11
6	Ne	20	90.92	21	0.06	22	8.82	20.18
7	Mg	24	78.6	25	10.11	26	11.29	24.33

Overall $A_r = (C4*D4+E4*F4+G4*H4)/(D4+F4+H4)$

3. a 84, 86, 83, 82, 80, 78
   b 84
   c Kr^{2+} ion
4. a 79, 81 are charged atoms, while 158, 160, 162 are molecular ions
   b Br-79 and Br-81 occur in equal proportions so are twice as likely to form 'mixed' molecules

5 Electron energy levels
1. Drawings equivalent to 2, 8, 8, 2 and 2, 8, 8
2. Drawings equivalent to:
   a 2, 8, 7
   b 2, 8, 8
   c 2, 8
   d 2, 8
   e 2, 8, 4
3. Level 1: 2s level 2: 2s,6p level 3: 2s,6p level 4: 2s
4.

Principal energy level	Sub-level s	Sub-level p	Sub-level d	Sub-level f	Total in principal energy level
1	2	x	x	x	2
2	2	6	x	x	8
3	2	6	10	x	18
4	2	6	10	14	32

5. a Groups I-II, s Groups III-VIII, p transition, d
   b Match equivalent sub-levels
   c 14
6. a Group 0
   b Group I
   c Group VII

6 Filling the levels
1. $3p^6$: 3 means principal energy level, p means p sub-level, 6 means six electrons
2. O $1s^2 2s^2 2p^4$
   Ne $1s^2 2s^2 2p^6$
   Na $1s^2 2s^2 2p^6 3s^1$
   Cl $1s^2 2s^2 2p^6 3s^2 3p^5$
   K $1s^2 2s^2 2p^6 3s^2 3p^6 4s^1$

Answers

3 Ti $1s^2 2s^2 2p^6 3s^2 3p^6 3d^2 4s^2$
   Mn $1s^2 2s^2 2p^6 3s^2 3p^6 3d^5 4s^2$
   Co $1s^2 2s^2 2p^6 3s^2 3p^6 3d^7 4s^2$
4 K [Ar] $4s^1$
   V [Ar] $3d^3 4s^2$
   Ni [Ar] $3d^8 4s^2$
   As [Ar] $3d^{10} 4s^2 4p^3$
   Br [Ar] $3d^{10} 4s^2 4p^5$
5 a S
   b Mg
   c Fe
   d Sr
   e Zr
   f Ba
6 a d shell filling before previous s shell (1)
   b Happens at half and full d shells
7 Co $1s^2 2s^2 2p^6 3s^2 3p^6 3d^{10} 4s^2 4p^6 4d^{10} 5s^1$

7 Electronic structure

1

Energy sub-level	Maximum number of electrons	Number of orbitals
s	2	1
p	6	3
d	10	5
f	14	7

2

Principal energy	Maximum number of electrons	Number of orbitals
1	2	1
2	8	4
3	18	9
4	32	16

3 Level number squared
4 Diagrams following the rule of singles first, then fill each level, e.g. O 2p 1 pair, two singles
5 Diagrams as for question 4
6 a Mn $1s^2\ 2s^2\ 2p^6\ 3s^2\ 3p^6\ 3d^5\ 4s^2$
      Mn^{2+} $1s^2\ 2s^2\ 2p^6\ 3s^2\ 3p^6\ 3d^5$
      Mn^{3+} $1s^2\ 2s^2\ 2p^6\ 3s^2\ 3p^6\ 3d^4$
   b Mn^{2+} as it has a half-full d shell
7 a Cu^{2+} in practice
   b Cu $1s^2\ 2s^2\ 2p^6\ 3s^2\ 3p^6\ 3d^{10}\ 4s^1$
      Cu^+ $1s^2\ 2s^2\ 2p^6\ 3s^2\ 3p^6\ 3d^{10}$
      Cu^{2+} $1s^2\ 2s^2\ 2p^6\ 3s^2\ 3p^6\ 3d^9$
      You would expect Cu^+ to be more stable as this has a full d-shell

8 Evidence for energy shells
1 a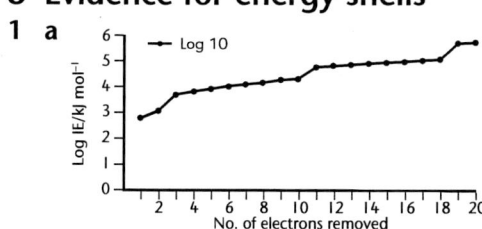
   b Four discrete zones clearly visible

   c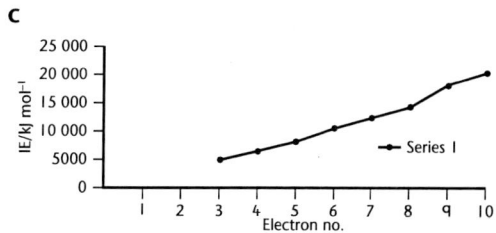

2 a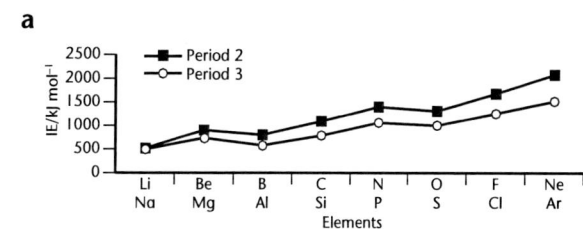

   b i   Change from s to p shell
      ii   First paired electron in p
      iii Period 3 shells further from nucleus
3 Sr – same shell, greater charge on nucleus
   Sb – as fewer principal levels
   Sc – same shell, greater charge on nucleus
   As – as p shell is half-full (unpaired)

9 Trends down a Group
1 a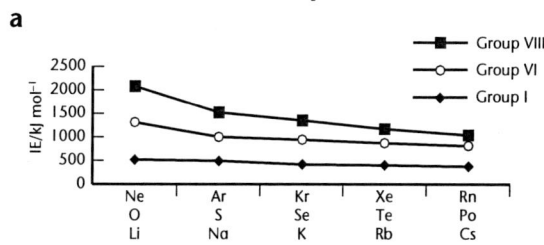

   b Rb just under 400, Te approx. 870, Kr approx. 1350
   c More levels – outer electron further from nucleus
   d Same level – greater charge
2 a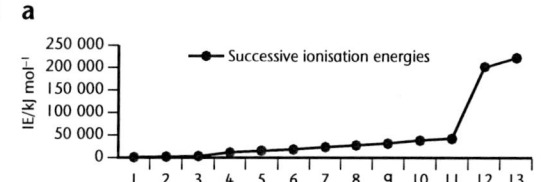

   b Breaks clearly visible – suitable annotations
   c 2p–2s not clear

   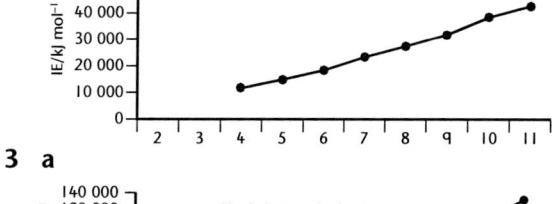

3 a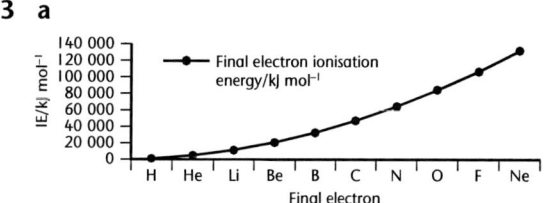

b Last electron always 1s – but charge gets progressively larger
c Na approx. 160 000

10 Ionic and covalent bonding
1 See text book
2 a CsF
  b K_2O
  c CaO
  d Na_3N
  e AlF_3
  f MgS
3 a Ionic
  b Covalent
  c Polar covalent (I δ+ Cl δ–)
  d 3 polar covalent N—H; 1 co-ordinate N—H using lone pair on N; ionic NH_4^+ Cl^-
  e Polar covalent (H δ^+ F δ^-)
  f Covalent: Al^{3+} has a high polarising power
  g 2 polar covalent O–H; 1 co-ordinate O–H using lone pair on O
  h Ionic
  i Covalent: Be^{2+} has a high polarising polar
  j Ionic with significant covalent character due to the high polarising power of Li^+
4 a MgF_2
  b MgO: Mg^{2+} gives rise to stronger ionic forces of attraction
  c NaCl is ionic whereas $AlCl_3$ is covalent
5 a Delocalised electrons
  b Smaller ions
  c Larger ions, lower charge and fewer delocalised electrons
  d Dipole–dipole attractions in IBr
  e Hydrogen bonding in HF
  f Weak van der Waals' forces between molecules, strong covalent bond in molecule
  g Ions must be mobile
  h Reacts by forming a co-ordinate bond
  i Weak van der Waals' forces between molecules
  j Does not form covalent bonds
  k One is co-ordinate as $AlCl_3$ can accept an electron pair

11 States of matter (1)
1 a Hydrogen bonding in ethanol, van der Waals' forces in propane
  b Ethylamine higher because of hydrogen bonding
2 a Hydrogen bonding
  b Dipole–dipole
  c Van der Waals'
  d Van der Waals'

3 Bonds in oil compounds not polar so not able to form hydrogen bonds
4 a F   b A   c E   d C
  e D   f G   g B
  A – NH_3, B – C_2H_5OH, C – SiO_2, D – Ca,
  E – $BaCl_2$, F – CO, G – I_2

12 States of matter (2)
1 Evaporation occurs from the surface when those molecules have acquired enough energy by collisions to overcome the intermolecular forces; boiling occurs throughout the liquid when all the molecules have been given enough energy to overcome the intermolecular forces
2 a Lower it
  b Raise it
3 a i Ionic
    ii Hydrogen bonding
  b O—H bonds are polar; δ^+ oxygen attracts K^+, δ^+ hydrogen attracts Cl^-
4 All hydrogen bonds are overcome on vaporising; only some on melting
5 Cs^+ larger than Na^+ so can fit more chloride ions around it
6 a Delocalised electrons
  b Graphite has weak van der Waals' forces between the layers; all the carbon atoms in diamond are fully covalently bonded
  c Strong covalent bonds
7 a iii   b iv   c i
  d ii   e v   f vi

13 Shapes of molecules and ions
1 a True – two electron pairs
  b True – three electron pairs
  c False – 107° three bond pairs, one lone pair so tetrahedral; lone pair repels bond pairs
  d False – 105° lone pair–lone pair repulsion
  e False – five bond pairs, one lone pair so shape based on octahedron
  f True – four electron pairs
  g True – two bond pairs, two lone pairs
  h True – six electron pairs
  i False – two bond pairs, two lone pairs so V-shaped
  j False – four bond pairs, one lone pair so shape based on trigonal bipyramid
2 a V-shaped; two bond pairs, two lone pairs; 105°
  b Linear; two bond pairs; three lone pairs; 180°
  c Linear; two bond pairs, three lone pairs; 180°
  d Five bond pairs, one lone pair; shape based on octahedron; 90°, 180°

Answers

e Square planar; four bond pairs, two lone pairs; 90°
f Square planar; four bond pairs, two lone pairs, 90°
g Tetrahedral; four bond pairs; 109.5°
h Three bond pairs, two lone pairs; shaped based on trigonal bipyramid; would expect 120° but in fact is 90°, 180° (slightly lower due to lone pair-lone pair repulsion)
i V-shaped; two bond pairs, two lone pairs; 105°
j Linear; two bond pairs, three lone pairs; 180°

3 a V-shaped; linear
   b i Co-ordinate
     ii Empty orbital
     iii Reduced from 120° to 109.5°; four bond pairs so tetrahedral
     iv NH_3

4 a Two bridging Cl atoms each form a co-ordinate bond; lone pair on Cl donated to empty orbital in $AlCl^3$
   b Decreases from 120° to $109\frac{1}{2}$°C

14 Group II elements

2 Increases
3 More e⁻ shells/more inner shielding e⁻/more than compensates for increase in nuclear charge/outer e⁻ less strongly attracted
4 Values from graph
6 Decreases/larger atoms/more shielding e⁻/outer e⁻ less tightly bound/easier to remove
7 Decreases/larger atoms/more shielding e⁻/outer or bonding e⁻ less strongly attracted to nucleus
8 Decreases/ larger atoms/ more shielding e⁻/weaker attraction for shared e⁻/metallic bonds weaker
9 V small atom/ little shielding by 2s e⁻/outer e⁻ strongly attracted/Be^{2+} would have very high charge:size ratio/strong polarizing power/compounds covalent rather than ionic

15 Period 3 (1)

1 Value for P from graph
2 Left to right nuclear charge increases/constant shielding effect/electrons in same main energy level/outer electrons more strongly attracted/so atoms get smaller
4 Follow same pattern as Period 3/with each Period 2 element having a higher value than same Group member
5 e⁻ removed from 3p in Al but 3s in Mg/3p higher energy than 3s/outer electron easier to remove/despite increase in nuclear charge
6 In P e⁻ removed from singly occupied 3p orbital/ but in S it is from one with 2e⁻/pair repel each other/easier to remove from shared 3p
7 3s and 3p
8 Sub-shells filled in order of increasing energy/4s lower energy than 3d/so 4s filled first/Period determined by number of main levels occupied
9 Power of attracting e⁻ in a covalent bond
10 Same shielding effect/e⁻ in same main energy level/nuclear charge increasing so bonding e⁻ more strongly attracted
11 Does not form covalent bonds

16 Compounds of Group II elements (1)

1 Graph must be sketched not plotted
2 1.14×10^{-5}
3 1.733×10^{-8}
4 Testing for sulphate ions in solution
5 a i To ensure as much $BaSO_4$ as possible precipitates/volume of water kept to minimum
     ii To produce as much $BaSO_4$ as possible
     iii To keep volume of water low/as much $BaSO_4$ as possible precipitated/$BaCl_2$ a source of soluble Ba^{2+}
     iv Some $BaSO_4$ will dissolve/keep to minimum
     v Lower solubility at lower temperature/less lost
   b Increase mass of Na_2SO_4/use more soluble Ba^{2+} salt/lower temperature/boil with $(NH_4)_2CO_3$/finer mesh of filter paper/repeat and work out average value
   c 5.922 g
   d Only 95.3% pure according to practical values – argue either way considering percentage errors in procedure and percentage of precipitate reclaimed

17 Compounds of Group II elements (2)

1 $M(s) + 2H_2O(l) \rightarrow M(OH)_2(aq) + H_2(g)$
2 Be – allow Mg if explained
3 Increases/atoms get larger/easier to donate outer e⁻/lower Ea not ΔH related
4 a First column: all $M(OH)_2$
     Second column: 43/58/74/122/171
     Fourth column: $-/2.07 \times 10^{-4}/1.62 \times 10^{-2}/8.20 \times 10^{-2}/2.75 \times 10^{-1}$
   b Increases

18 Period 3 (2)

1 Na, Mg, Al metallic/Si macromolecular/P, S, Cl covalent/Ar atomic
2 Metals

Answers

3 Delocalised or mobile electrons

5 Metallic bond strength depends on number of electrons shared/size of the atoms
Macromolecular – strong covalent bonds must be broken/very high melting point
Molecular – only van der Waals' forces or induced dipole-dipole interactions/depend on molecular size/$P_4 < S_8 > Cl_2 < P_4$/Ar atomic – very weak van der Waals' forces

7 For melting some of the attractive forces must be overcome/for boiling all must be overcome

8 To melt Si many strong/covalent bonds must be broken/Al atoms small/good e- attractors/ so atoms still strongly attracted to each other when molten

19 Relative atomic and molecular mass

1 18.99
2 a 28.1
  b 144.2
3 197.5
4 102.91, rhodium
5 $X = 23.0$ (Na), $Y = 70.0$ (Ga)
6 Avogadro's constant × mass of a single atom
7 5.87×10^{20} molecules of bromine

20 Avogadro's constant and moles (1)

1 3.614×10^{25} atoms of carbon
2 2.15×10^{17} atoms of toxic metal
3 9.64×10^{22} nitrate ions
4 a 129 g of sodium
  b 269 dm³ of chlorine
5 75.3%
6 a 5950 cm³
  b 4824 g of C
7 33.00

21 Avogadro's constant and moles (2)

1 a F   b T   c T   d F
  e F   f T   g F
2 0.0214 mol
3 BaO_2
4 a NH_2
  b M_r

22 Empirical and molecular formulae

1 C_5H_4, $C_{10}H_8$
2 $C_4H_8O_2$
3 a $CoSO_8H_{10}$
  b 5 moles of water
  c SO_3^{2-}
4 a $NaHCO_3$
  b $C_3H_4O_3$
  c $NaHCO_3$ ($M_r = 84$), $C_6H_8O_6$ ($M_r = 176$)

5 a Charge not balanced
  b Incorrect formulae (RbOH, RbBr)
  c Incorrect state symbols (Mg(s), H_2(g))
  d Number of atoms not balanced on each side of equation

23 Balancing equations

1 a ICl_3
  b IF_4^+
  c $2Cl^-$
2 a $3NH_3 + H_3PO_4 \rightarrow (NH_4)_3PO_4$
  b $Li_2CO_3 \rightarrow Li_2O + CO_2$
  c $2Cu^+ \rightarrow Cu^{2+} + Cu$
  d $P_4 + 3OH^- + 3H_2O \rightarrow 3H_2PO_2^- + PH_3$
  e $2Fe^{2+} + Cl_2 \rightarrow 2Fe^{3+} + 2Cl^-$
  f $4HgS + 4CaO \rightarrow 4Hg + CaSO_4 + 3CaS$
  g $2ND_3 + 3CuO \rightarrow N_2 + 3Cu + 3D_2O$
  h $5CO + I_2O_5 \rightarrow 5CO_2 + I_2$
3 a $Sr + 2H_2O \rightarrow Sr(OH)_2 + H_2$
  b $KrF_2 \rightarrow F_2 + Kr$
  c $Al_4C_3 + 12H_2O \rightarrow 3CH_4 + 4Al(OH)_3$
  d $C_{12}H_{22}O_{11} + H_2O \rightarrow 4CO_2 + 4C_2H_5OH$
4 a $SO_2 + V_2O_5 \rightarrow V_2O_4 + SO_3$, $V_2O_4 + \tfrac{1}{2}O_2 \rightarrow V_2O_5$
  b $SO_2 + \tfrac{1}{2}O_2 \rightleftharpoons SO_3$
5 a C_2^{2-}
  b U^{6+}
  c $IrCl_6^{3-}$
  d $[Be(OH)_4]^{2-}$
  e Pu^{4+}

24 Reacting masses

1 6.6 g S
2 68.9 g Cu
3 816 kg Al_2Cl_6, 0.06 kg Al
4 a Sulphuric acid
  b 20 g
  c 60%

25 Reacting masses and volumes

1 a 20.58 g
  b 24.55 g
  c 25 cm³
2 For two gases $P_1V_1 = n_1RT_1$ and $P_2V_2 = n_2RT_2$. If both gases are at the same T and P then $P_1/T_1 = P_2/T_2$ and R is constant so $n_1/V_1 = n_2/V_2$. If $V_1 = V_2$ then n_1 must $= n_2$
3 0.5 dm³
4 Yes

26 Reacting volumes

1 12.5 dm³
2 66.0 dm³ of ammonia and 49.5 dm³ of oxygen
3 $N_2F_4 + Cl_2 \rightarrow 2NF_2Cl$
4 C_3H_8

Answers

5 $1100 \, cm^3$
6 $4.8 \, dm^3$
7 $5.50 \, g$

27 The ideal gas equation (1)
1 3008
2 $9.434 \, m^3$
3 $33.8 \, g$
4 $3330 \, Pa \, m^{-3}$ or J
5 $1.42 \, g \, dm^{-3}$
6 $5482 \, Pa$
7 44.00
8 ECl_4

28 The ideal gas equation (2)
1 a 121.0
   b CF_2Cl_2, CF_2Cl_2
2 a $C_3H_5N_3O_9 \rightarrow 3CO_2 + 2.5H_2O + 1.5N_2 + 0.25O_2$
   b 0.508 mol
   c 1090 kPa
3 a $2NaN_3 \rightarrow 2Na + 3N_2$, $10Na + 2KNO_3 \rightarrow K_2O + 5Na_2O + N_2$
   b 166 g NaN_3, 51.5 g KNO_3

29 Molarity and solutions
1 a 0.078 M
   b 0.0064 M
   c 0.0060 M
2 a $40 \, g \, dm^{-3}$
   b $5.38 \, g \, dm^{-3}$
   c $0.342 \, g \, dm^{-3}$
   d $315 \, g \, dm^{-3}$
3 $M_r = 42$
   a 50.4 g
   b 0.063 g
   c 12.6 g
4 $242 \, cm^3$
5 $6.87 \, mol \, dm^{-3}$
6 a $1200 \, cm^3$
   b $6000 \, cm^3$
7 $429 \, cm^3$

30 Volumetric analysis (1)
1 a Indicator, end point
   b $Ba(OH)_2 + 2HCl \rightarrow BaCl_2 + 2H_2O$, $0.0447 \, mol \, dm^{-3}$, $7.64 \, g \, dm^{-3}$
2 42.43%
3 2.12 g NaOH, 14.35 g RbOH

31 Volumetric analysis (2)
1 $GeBr_4 + 4H_2O \rightarrow Ge(OH)_4 + 4HBr$, Ge
2 79.1%
3 $KH_3C_4O_8 \cdot 2H_2O$

32 Enthalpy change and calorimetry
1 See text book
2 a $H_2(g) + \frac{1}{2}O_2(g) \rightarrow H_2O(l)$
   b $C(s) + 2H_2(g) + \frac{1}{2}O_2(g) \rightarrow CH_3OH(l)$
   c $2C(s) + 3H_2(g) + \frac{1}{2}O_2(g) \rightarrow C_2H_5OH(l)$
   d $Al(s) + 1\frac{1}{2}Cl_2(g) \rightarrow AlCl_3(s)$
   e $2Na(s) + C(s) + 1\frac{1}{2}O_2(g) \rightarrow Na_2CO_3(s)$
   f $Cu(s) + S(s) + 5H_2(g) + 4\frac{1}{2}O_2(g) \rightarrow CuSO_4 \cdot 5H_2O(s)$
3 a $CH_4(g) + 2O_2(g) \rightarrow CO_2(g) + 2H_2O(l)$
   b $C_4H_{10}(g) + 6\frac{1}{2}O_2(g) \rightarrow 4CO_2(g) + 5H_2O(l)$
   c $CH_3OH(l) + 1\frac{1}{2}O_2(g) \rightarrow CO_2(g) + 2H_2O(l)$
   d $C_2H_5OH(l) + 3O_2(g) \rightarrow 2CO_2(g) + 3H_2O(l)$
   e $CH_3COOH(l) + 2O_2(g) \rightarrow 2CO_2(g) + 2H_2O(l)$
4 a $-1560 \, kJ \, mol^{-1}$
   b $-2218 \, kJ \, mol^{-1}$
   c $-3514 \, kJ \, mol^{-1}$
5 a 520.8 J
   b 0.00217
   c $-239.6 \, kJ \, mol^{-1}$
   d More time for heat to be lost
6 a 5712 J
   b $-57.1 \, kJ \, mol^{-1}$

33 Calorimetry and using Hess's Law
1 a 882 J
   b 0.0594
   c $+14.8 \, kJ \, mol^{-1}$
2 a 2074.8 J
   b 0.004
   c $-518.7 \, kJ \, mol^{-1}$
3 13 720 J, ethanol
4 a 0.302 g
   b 49.6°
5 a $-528 \, kJ \, mol^{-1}$
   b $-72 \, kJ \, mol^{-1}$
6 $-54 \, kJ \, mol^{-1}$

34 Calculations using Hess's Law
1 $+440 \, kJ \, mol^{-1}$
2 a $-76 \, kJ \, mol^{-1}$ $-49 \, kJ \, mol^{-1}$ $-2881 \, kJ \, mol^{-1}$ $-251 \, kJ \, mol^{-1}$ $-1368 \, kJ \, mol^{-1}$ $-484 \, kJ \, mol^{-1}$ $-2239 \, kJ \, mol^{-1}$
   ΔH for the reaction is $-5 \, kJ \, mol^{-1}$
3 a $-2046 \, kJ \, mol^{-1}$
   b $-2056 \, kJ \, mol^{-1}$
   c Bond enthalpies are mean values
4 a $-76 \, kJ \, mol^{-1}$
   b $-11 \, kJ \, mol^{-1}$
   c $-242 \, kJ \, mol^{-1}$
5 $338 \, kJ \, mol^{-1}$

Answers

35 Thermochemical calculations

1. a $-39.8\,kJ\,mol^{-1}$, $+24.8\,kJ\,mol^{-1}$
   b $+89.4\,kJ\,mol^{-1}$
   c Not enough HCl to react
2. i a $-72.4\,kJ\,mol^{-1}$ b $+5.01\,kJ\,mol^{-1}$
   ii $-77.4\,kJ\,mol^{-1}$
3. a $-541\,kJ\,mol^{-1}$
   b $180.3\,kJ$
   c $62.9\,°C$
   d i $-847\,kJ\,mol^{-1}$
       ii $21.4\,g$
       iii No; extra energy would be needed to overcome the intermolecular forces

36 Collision theory and rate of reaction (1)

1. a Twigs have a larger surface area than logs, so oxygen comes into contact with more of the wood. Therefore there are more fruitful collisions between molecules of oxygen and wood.
   b There are fewer hydrogen ions in an equal volume of diluted vinegar. Therefore there are fewer fruitful collisions with the bicarbonate of soda.
   c Lowering pressure decreases the concentration of the gases. Therefore there are fewer fruitful collisions between freely moving molecules. In the solid, molecules can only vibrate and are closely packed so collisions between molecules are not influenced by pressure.
   d Iron powder is a catalyst. The catalyst lowers the activation energy of the reaction by providing an alternative pathway for the reaction. Therefore more collisions will be fruitful.
   e In the solid molecules can only vibrate, so there are very few fruitful collisions.
   f At a higher temperature many more molecules have reached the activation energy, so there are more fruitful collisions.
   g Making the vessel larger decreases the concentration of the gases. Therefore there are a smaller number of molecules in a given unit of volume, so fewer fruitful collisions.
2. a O depends on geometry and alignment of molecules during fruitful collision.
       C depends on how closely the molecules are crowded together, i.e. concentration/pressure and also mass of molecules × temperature.
       E depends on the fraction of molecules with the E_a and/or the magnitude of E_a.
   b E a small increase in temperature causes a large increase in the number of molecules having E_a (exponential relationship). At a fixed temperature a small change in E_a has a large effect on the fraction of fruitful energetic collisions.

37 Collision theory and rate of reaction (2)

1. a Record the time taken to obscure a cross marked on paper and placed below the reaction vessel.
   b Temperature influences the rate of a reaction, so this variable must be held constant in the investigation.
   c Water is added to ensure the concentration of hydrogen ions remains constant as the rate of reaction is inflenced by the concentration of this reactant.
   d Rate decreased. The concentration of the acid is lowered by dilution. Therefore fewer hydrogen ions per unit of volume of solution, so fewer fruitful collisions.
   e Graph should have two sensible scales, points plotted correctly, straight line (best fit) through the origin.
   f Value from graph.
   g Rate of the reaction is proportional to the concentration of thiosulphate ions.
      $R = k \times$ [thiosulphate ions]
   h At a higher concentration there are more thiosulphate ions per unit of volume of solution, so there are more fruitful collisions. Rate depends on (fruitful) collision frequency.
   i Straight line which passes through the origin but gradient is steeper. At any given concentration of thiosulphate ions the reaction rate is quicker at a higher temperature. At the higher temperature many more ions have the activation energy.

38 Factors influencing reaction rate (1)

1. a Hydrogen
   b $M + 2HBr \rightarrow MBr_2 + H_2$ or
      $M + 2H^+ \rightarrow M^{2+} + H_2$
   c d-block (transition elements)
   d Monitor loss in mass (use a balance). Monitor increase in the intensity of colour of the reaction mixture (use a colorimeter).

39 Factors influencing reaction rate (2)

1. a Graph should have two sensible scales, points plotted correctly, smooth curve.
   b Value from graph
   c Value from graph

Answers

d 1st minute 18 cm³, 2nd minute 10 cm³, 3rd minute 5 cm³. Rate of reaction is greatest at the begining of the reaction. Rate goes down progressively with time. At the start of the reaction the number of hydrogen ions is at a maximum, as the ions react with the metal the number of fruitful collisions decreases.

e Same shape curve but to the right (less steep gradient). Starts at 0 and flattens at the same final volume of gas. Less surface area, chances of fruitful collisions reduced, rate decreased, longer time for acid to be used up. Same final volume of gas when reaction is complete because an excess of metal is used with the same volume and concentration of acid.

f Same shape curve as original but to the left (steeper gradient). Starts at 0 and flattens at same final volume of gas. Reaction stops in a shorter time. At a higher temperature many more ions have the activation enerygy to react. Therefore there are more fruitful collisions in a given unit of time, so the reaction is faster. Same final volume of gas when reaction is complete because an excess of metal is used with the same volume and concentration of acid.

g 64 times as fast

h 36 cm³ of hydrogen gas, 0.003 g of hydrogen gas

i i $0.333 \, cm^3 s^{-1}$
   ii $0.0000139 \, mol \, s^{-1}$

j $0.000463 \, mol \, dm^{-3} s^{-1}$

k i $0.00000694 \, mol \, s^{-1}$
   ii $0.00000278 \, mol \, s^{-1}$

l Ethanoic acid is a weak acid. For an equal volume of hydrobromic acid there are fewer hydrogen ions. Therefore there are fewer fruitful collisions between the ions and the metal.

40 Maxwell–Boltzmann distribution

1 a X: energy, Y: number or fraction of molecules with a given energy

   b All molecular collisions do not result in a chemical change. Only a small proportion of molecules react on collision. These molecules react because they possess the activation energy (the minimum energy needed for a reaction to occcur).

   c Number of molecules with energy less than the activation energy

   d t_b: similar shape curve but higher and to the left of curve for t_a. At t_b fewer molecules have E_a (smaller shaded area). Therefore fewer fruitful collisions.

   e Yes, same number of molecules at t_a and t_b

   f Add a catalyst. Catalyst provides an alternative reaction pathway of lower activation energy. Therefore more molecules have sufficient energy to react.

   g The area under the curve would be reduced (fewer molecules present) and the hump of the curve would be lower but the most probable energy is at the same value.

2 b E_a (back reaction) = $173 \, kJ \, mol^{-1}$
   c E_a (catalysed reaction) = $71.3 \, kJ \, mol^{-1}$

41 Catalysts

3 d-block elements

4 A catalyst provides an alternative pathway of lower activation energy. Therefore more molecules have sufficient energy to react, so more fruitful collisions. One example of an homogeneous catalyst is concentrated sulphuric acid which can be used in the dehydrolysis of alcohols.

5 A homogeneous catalyst is in same physical state as the reactants and products. A heterogeneous catalyst is in a different physical state to the reactants and products. A homogeneous catalysed reaction proceeds through intermediate species formed in several reactions. These reactions are much quicker (have lower values of E_a) than the uncatalysed reaction. Heterogeneous catalysis often takes place at a surface at active sites. Adsorption of reactants to the surface weakens chemical bonds. Products desorb to regenerate active sites.

6 Enzymes are highly specific in their action and are enormously efficient. Enzymes function in mild conditions found within living cells and can operate at room temperature and pressure.

42 Establishing equilibria

1 Reaction goes both ways or is reversible.

2 Dynamic: reaction still going on/rates not zero
   Equilibrium: forward rate = reverse rate

3 a equal
   b stay constant – not "the same"

4 Initially forward rate unchanged as [N_2], and [H_2] not changed/slows down as N_2 and H_2 converted to NH_3. Reverse rate goes down initially as [NH_3] lower/increases as net formation of NH_3 occurs. Eventually new equilibrium established with both rates lower than original as all concentrations lower than at start.

© Farrow R, Gibbens D, Stirrup M and Vowles R, 2000. AS Chemistry for AQA – Resource Pack, Heinemann

Answers

5 b Read off time when graph first becomes horizontal

c Start at 0, increase to 0.672, mirror plotted line

6 $CaCO_3(s) + 2HCl(aq) = CaCl_2(aq) + H_2O(l) + CO_2(g)$

CO_2 diffuses away/no reverse reaction/if reverse rate 0, equilibrium only when forward rate also 0, i.e. when reaction complete

43 Applying Le Chatelier's Principle

1 Endothermic forward reaction/higher temperature, more NO_2 or less N_2O_4/less brown

2 Darken/increase in pressure favours fewer moles of gas/more N_2O_4 formed/more brown

3 b Start at 0/line going through $2(1-n)$ where n = moles of N_2O_4/levels out at 1.44 mol

c 0.28/equilibrium reached before 25 min

4 No change as long as volume stays the same

44 A gaseous equilibrium

1 a Exothermic/higher yield at lower temperature /at any given pressure (except 0)

b Decreases/more NH_3 as pressure goes up/higher pressure favours reaction giving fewer moles of gas

c i yield up
ii yield up
iii no change

2 High pressure/decrease in moles of gas from left to right

Moderate temperature/lower temperature gives higher yield but slower rate

Catalyst/increases rate without affecting yield

45 Production of ethanol

1 Higher pressure gives higher yield as number of moles of gas decreases from left to right

2 Costs go up more quickly than yield as pressure increased/eventually becomes uneconomic to increase pressure further

3 Yield down/reaction exothermic

4 Lower temperature gives slower reaction

5 Rate increased/without decreasing yield

6 Gives large surface area/helps to keep catalyst in reaction chamber

7 Easy to remove ethanol and recycle unreacted ethene

8 a Lower labour costs/large amounts of ethanol produced quickly

b Large demand/more profit or cheaper product/more competitive

9 Process important in imparting flavour

46 Production of ammonia

1 Methane or natural gas/air/water

2 a Both go to right/both exothermic

b reaction (1) goes to right/more moles of gas on right, reaction (2) no change/moles of gas equal on left and right

3 Low temperature for both/low pressure for reaction (1)/reaction (2) not influenced by pressure so pressure irrelevant

4 Lower temperature gives higher yield/lower fuel costs/but slower reaction. A catalyst/increases rate/without decreasing yield/cost of catalyst initially high but it is not consumed so economic in long term. Lower pressure cheaper/but reaction slower/better yield for reaction (1)/but no change for reaction (2)

5 Low temperature/ high pressure

6 ~600 K/35–60 MPa/iron catalyst

7 Temperature: moderate temperature used/better yield at lower temperature but reaction slower so a temperature is needed that will give an acceptable yield at reasonable rate. Pressure: higher pressure gives higher yield but at greater cost so pressure used that gives acceptable yield at reasonable cost. Catalyst: increases rate without decreasing yield or increasing plant running costs in the long term

8 Fertilizer production – not just "fertilizer"

47 Production of sulphuric acid

1 S/O_2 either way round/O_2/H_2SO_4/H_2O

2 Ovals: inputs or raw materials/boxes: intermediates and product

3 A: $S + O_2 = SO_2$
B: $2SO_2 + O_2 = 2SO_3$

4 Low temperature/both exothermic
High pressure/both have fewer moles of gas on right-hand side

5 Higher pressure would give a higher yield but at increased cost/yield very high already/uneconomic to increase it

6 Reactants acidic/acid rain/kills plants and animals/damages buildings

7 Need to import S/close to a port/on the coast/need to transport product/close to rail or motorway/near to customers/need a skilled work force/close to major city or town or other chemical industries/environmental/on east coast as prevailing wind from south-west

Answers

48 Oxidation and reduction
1. a removing oxygen
   b C
   c solid, liquid, gas
2. a Cu reduced, H oxidised
   b Fe reduced, Al oxidised
3. a potassium oxide formation, K oxidised, O reduced
   b sodium chloride formation, Na oxidised, Cl reduced
   c iron sulphide formation, Fe oxidised, S reduced
   d iron(III) oxide formation, Fe oxidised, O reduced
4. a–c metal oxidised to ion, metal ion in compound reduced to metal
   d–e halogen reduced to halide, halide in compound oxidised to halogen
   f–g metal oxidised to ion in salt, hydrogen in acid/water reduced to element

49 Oxidation states (1)
1. Charge on simple ion/elements zero/ compound sum to zero
2. a +3   b –2   c 0   d –1
   e +2   f +1   g 0
3. a +4   b +2   c +3   d +1
4. a –1   b –3   c –4   d –2
   e –2   f –3
5. K +1, Ca +2, Na +1, Al +3, F –1, O –2, H +1
6. Li 1, S –2, Br –1, Fe +3, Cl –1, P +3, Cl –1, P +5, O –2, Cr +6, O –2, Cr +3, O –2, Cl +4, Ca 2, C –1
7. a H –1
   b O –1

50 Oxidation states (2)
1. P +5
2. a S +6
   b I +5
   c N +5
   d Cr +6
   e Cr +6
3. a 4   b 5   c 6   d 7
   Group number
4. a –1   b –1   c –2   d –3
5. a +7   b +5   c +3

51 What's in a name?
1. a Copper(II) sulphate
   b Lead(II) sulphide
   c Manganese(IV) oxide
   d Chromium(III) sulphate
   e Iron(II) sulphate
2. Iron(II) chloride, iron(III) chloride
3. a $CoCl_2$
   b $AgNO_3$
   c $ZnCO_3$
   d PbO
   e PbO_2
   f $SbCl_5$
4. a +3   b +7   c +5
5. -ide means just Cl, i.e. NaCl; -ate means with oxygen, sodium chlorate(I) is NaClO
6. Na_2SO_3, sodium sulphate(IV), KClO, potassium chlorate(I), Ca $(NO_2)_2$, calcium nitrate(III), $Fe(OH)_2$, iron(II)hydroxide

52 Oxidation states and redox
1. a Decreases
   b Increases
   c Rises with oxidation, falls with reduction
2. a Zn up 2 so oxidised, Cu down 2 so reduced
   b same answer as a
3. $Br_2 + 2KI \rightarrow I_2 + 2KBr$; $Br_2 + 2I^- \rightarrow I_2 + 2Br^-$; Br down so reduced, I up so oxidised
4. S from 4 to 6 so oxidised; I from 5 to 0 so reduced
5. S from +4 to +6 so oxidised; Br from 0 to –1 so reduced
6. Reduction, Fe +3 to +2 so reduction, I from –1 to 0 so oxidation; iodine reduces the Fe but is itself oxidised hence link to loss of electrons
7. Cr +6 to +3 so is itself reduced as it oxidises the alcohol – Cr gains electrons

53 Simple half-equations
1. a Oxidation
   b Oxidation
   c Reduction
   d Reduction
2. a $Pb^{4+} + 2e^- \rightarrow Pb^{2+}$
   b $Ba \rightarrow Ba^{2+} + 2e^-$
   c $2F^- \rightarrow F_2 + 2e^-$
   d $I_2 + 2e^- \rightarrow 2I^-$
3.

Write down the formulae and balance the atoms undergoing redox	[HI $\rightarrow I_2$] unbalanced... $2HI \rightarrow I_2$
Balance any oxygen atoms present by adding water	Not required
Balance any hydrogen atoms by adding H^+ ions	$2HI \rightarrow I_2 + 2H^+$
Balance the charges by adding electrons	$2HI \rightarrow I_2 + 2H^+ + 2e^-$

4.

Write down the formulae and balance the atoms undergoing redox	$MnO_4^- \rightarrow Mn^{2+}$
Balance any oxygen atoms present by adding water	$MnO_4^- \rightarrow Mn^{2+} + 4H_2O$
Balance any hydrogen atoms by adding H^+ ions	$MnO_4^- + 8H^+ \rightarrow Mn^{2+} + 4H_2O$
Balance the charges by adding electrons	$MnO_4^- + 8H^+ + 5e^- \rightarrow Mn^{2+} + 4H_2O$
Check that the number of atoms is balanced, the charges are balanced, and that the number of electrons transferred is equal to the change in oxidation number	

Answers

5

Write down the formulae and balance the atoms undergoing redox	$ClO^- \rightarrow Cl^-$
Balance any oxygen atoms present by adding water	$ClO^- \rightarrow Cl^- + H_2O$
Balance any hydrogen atoms by adding H^+ ions	$ClO^- + 2H^+ \rightarrow Cl^- + H_2O$
Balance the charges by adding electrons	$ClO^- + 2H^+ + 2e^- \rightarrow Cl^- + H_2O$
Check that the number of atoms is balanced, the charges are balanced, and that the total number of electrons transferred is equal to the change in oxidation number	$ClO^-(aq) + 2H^+(aq) + 2e^- \rightarrow Cl^-(aq) + H_2O(l)$

54 Constructing whole equations

1

The half-equation for the oxidation part	$ClO^- + 2H^+ + 2e^- \rightarrow Cl^- + H_2O$
The half-equation for the reduction part	$2I^- \rightarrow I_2 + 2e^-$
Multiply up to match electrons (if necessary)	
Add the two together	$ClO^- + 2H^+ + 2e^- + 2I^- \rightarrow Cl^- + H_2O + I_2 + 2e^-$
Cancel out the electrons	$ClO^- + 2H^+ + 2I^- \rightarrow Cl^- + H_2O + I_2$
Check that atoms and charges balance, add state symbols	$ClO^-(aq) + 2H^+(aq) + 2I^- \rightarrow Cl^-(aq) + H_2O(l) + I_2(s)(or\ aq)$

2

The half-equation for the oxidation part	$Fe^{2+} \rightarrow Fe^{3+} + e^-$
The half-equation for the reduction part	$MnO_4^- + 8H^+ + 5e^- \rightarrow Mn^{2+} + 4H_2O$
Multiply up to match electrons	$5Fe^{2+} \rightarrow 5Fe^{3+} + 5e^-$
Add the two together	$MnO_4^- + 8H^+ + 5e^- + 5Fe^{2+} \rightarrow Mn^{2+} + 4H_2O + 5Fe^{3+} + 5e^-$
Cancel out the electrons	$MnO_4^- + 8H^+ + 5Fe^{2+} \rightarrow Mn^{2+} + 4H_2O + 5Fe^{3+}$
Check that atoms and charges balance, add state symbols	$MnO_4^-(aq) + 8H^+(aq) + 5Fe^{2+}(aq) \rightarrow Mn^{2+}(aq) + 4H_2O(l) + 5Fe^{3+}(aq)$

3

The half-equation for the oxidation part	$Br^- \rightarrow \frac{1}{2}Br_2 + e^-$
The half-equation for the reduction part	$SO_4^{2-} + 4H^+ + 2e^- \rightarrow SO_2 + 2H_2O$
Multiply up to match electrons	$2Br^- \rightarrow Br_2 + 2e^-$
Add the two together	$SO_4^{2-} + 4H^+ + 2Br^- + 2e^- \rightarrow SO_2 + 2H_2O + Br_2 + 2e^-$
Cancel out the electrons	$SO_4^{2-} + 4H^+ + 2Br^- \rightarrow SO_2 + 2H_2O + Br_2$
Check that atoms and charges balance, add state symbols	$SO_4^{2-}(aq) + 4H^+(aq) + 2Br^-(aq) \rightarrow SO_2(aq) + 2H_2O(l) + Br_2(aq)$

55 Trends

1 a

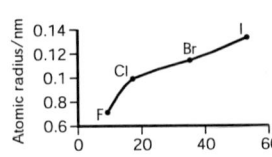
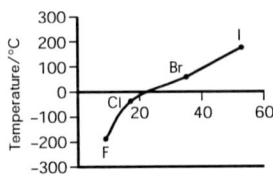
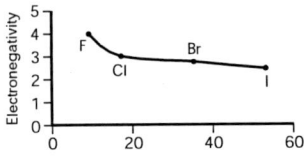

b 2, 3, 4, 5 shells (energy levels)
Van der Waals' forces linked to atom size
Bigger atoms attract outer, bonding electrons, less strongly

c F–Cl no d orbital; Cl–Br, Br–I have d-orbital between them

2 a Approx. 0.15 nm, approx. 320 °C, approx. 2.2

b Atomic radius

c Similar – noble gas slightly smaller

d Radon, 0.145 so astatine a little bigger at approx. 0.15, fairly reliable

e 337 °C – a little higher than prediction

3 F_2 strongest oxidising agent, as smallest atoms, therefore greatest ability to attract electrons. Converse for I_2

56 Halogens are oxidising agents

1 a Half equations:
$2K \rightarrow 2K^+ + 2e^-$ potassium loses electrons (OIL – oxidation)
$I_2 + 2e^- \rightarrow 2I^-$ iodine accepts electrons (RIG – reduction)

b Half equations:
$Mg \rightarrow Mg^{2+} + 2e^-$ magnesium loses electrons (OIL – oxidation)
$F_2 + 2e^- \rightarrow 2F^-$ fluorine accepts electrons (RIG – reduction)

c Half equations:
$2Fe^{2+} \rightarrow 2Fe^{3+} + 2e^-$ iron(II) ions lose electrons (OIL – oxidation)
$Br_2 + 2e^- \rightarrow 2Br^-$ bromine accepts electrons (RIG – reduction)
$2Cl^-$

2 a Greater oxidising power up the group

b, c $Cl_2 + 2e^- \rightarrow 2Cl^-$ red (RIG)
$[2K^+] + 2I^- \rightarrow I_2 + 2e^- + [2K^+]$ ox (OIL)

3

	Cl^- (aq)	Br^- (aq)	I^- (aq)
Chlorine in solvent	green	brown	purple
Bromine in solvent	brown	brown	purple
Iodine in solvent	purple	purple	purple

4 a Chlorine displaces bromine (oxidation), the bromine is reduced with SO_2 then re-oxidised with chlorine

b Bromine is blown out with air (forced evaporation), the bromide ions dissolve in water, distillation

57 Halides are reducing agents

1 a Half equations: $[Mg] + 2Br^- \rightarrow Br_2 + 2e^-$
bromide ions lose electrons (OIL – oxidation)
$Cl_2 + 2e^- \rightarrow 2Cl^-$
chlorine accepts electrons (RIG – reduction)

Answers

b Half equations: $Cl^- \rightarrow \frac{1}{2}Cl_2 + e^-$
chloride ion loses electron (OIL – oxidation)
$Ag^+ + e^- \rightarrow Ag$
silver ion accepts electron (RIG – reduction)

c Half equations: $2Cl^- \rightarrow Cl_2 + 2e^-$
chloride ions lose electrons (OIL – oxidation)
$Mn^{4+} + 2e^- \rightarrow Mn^{2+}$
manganese(IV) ion accepts electrons (RIG – reduction)

2 a Increases down the group
  b Larger ions find it easier to loses electrons
3 A: Bromide (Br_2 fumes and SO_2), B: chloride HCl only, C: iodide (I_2 fumes and H_2S)
4

Reaction	Oxidation number of the sulphur atoms		Is the sulphur reduced?
	Reactants	Products	
A	6	6	no
B	6	4	yes
C	6	–2	yes

Increased reducing power down the group: change in oxidation number: 0, –2, –8

58 Which halide?
1 a, b

Halide ions (aq)	With $AgNO_3$(aq)	Plus dilute NH_3(aq)	Plus concentrated NH_3(aq)	With $Pb(NO_3)_2$(aq) (see part b)
F^-	No reaction	No reaction	No reaction	White precipitate
Cl^-	White precipitate	Dissolves	Dissolves	White precipitate
Br^-	Cream precipitate	No change	Precipitate dissolves	Yellow precipitate
I^-	Yellow precipitate	No change	No change	Yellow precipitate

c Lithium fluoride
d Magnesium bromide – will not dissolve in dilute ammonia
e Silver chloride breaks down to silver
f Sodium iodide

2 a i Bromide ions emit electrons when hit by light – these are accepted by the silver ions which are reduced
  ii Further reduction of silver ions occurs around the silver atoms from i
  b The rest of the silver bromide would break down when the film was exposed to light
  c Original is a negative, i.e. black with silver where light fell, so must repeat the process to get a positive image

59 Halogens in water
1 a 0, –1 and 1
  b Reduced
  c Oxidised
  d Bromine is both oxidised and reduced in the same reaction
2 a H^+ ions at first, then bromate(I) ion bleaches
  b H^+ ions removed so drives the reaction to the right to make up the shortfall – bromine removed
  c Excess of H^+ ions drives reaction to the left – more bromine
3 a Calcium Chlorate(I) and calcium chloride
  b $2Cl_2 + 2Ca(OH)_2 \rightarrow 2H_2O + CaCl_2 + Ca(ClO)_2$
  c Chlorine gas would be released
4 a Oxidation (OIL)
  b $H_2 + Cl_2 \rightarrow 2HCl$ (H ox, Cl red)
  c Cl_2 and NaOH are both primary products – combine to make bleach
  d Cl +1 goes to Cl –1 [× 2] plus Cl +5
5 Solvents, refrigerants, aerosol propellants, they damage the ozone layer

60 Value for money
1 a Sodium chlorate(I), chlorine
  b

1 A known volume of bleach is taken	Must know volume to calculate molarity
2 An excess of potassium iodide is added	So that *all* of the chlorine is used up
3 Sulphuric acid is added	Chlorine released from bleach – displaces the iodine
4 This mixture is titrated against sodium thiosulphate solution of know molarity until the yellow colour disappears	Yellow colour disappears when all the iodine has reacted
5 This is repeated, this time adding starch solution just before the end point	Better indicator of the end-point

c If acid was added first chlorine released may be lost – hazard
d The blue-black starch–iodine colour is easier to spot as it disappears

2

Name	Price/£	Bottle volume /cm^3	Calculated molarity	Mol pound^{-1}
Superthick bleach	1.75	1000	0.15	0.086
Supermarket best	2.12	2000	0.12	0.113
Cheapo bleach (new stock)	1.29	1136	0.09	0.079
Cheapo bleach (old stock)	1.29	1136	0.075	0.066

'Supermarket best' gives the best value for money.

Answers

3 a Number of moles of thiosulphate = volume × molarity/1000 = 0.000235
   But 2 moles of thiosulphate are needed for every 1 mole of iodine (hence chlorine) so:
   Number of moles of chlorine in the $25\,cm^3$ dilute bleach sample = 0.000118
   Number of moles in $500\,cm^3$ of dilute = no. of moles in $25\,cm^3$ of original bleach = 0.00235
   Therefore the concentration of chlorine in the original bleach (mol/dm^{-3}) = 0.094
 b Probably 'cheapo' bleach (fairly new stock)

61 Principles of extraction

1 a O, Si
  b Al, Fe
  c Good conductors of heat and electricity, malleable, ductile, etc.
2 They occur in high grade ores in a few specific locations.
3 a Reduction
  b Cost of the reducing agent, cost of the energy, required purity of the metal
  c i Cheap or abundant
    ii CO toxic; CO_2 greenhouse gas
  d i Only small amounts required
    ii Pure product, no pollution; risk of explosion, storage/transport of large volumes of gas
  e i Very reactive metal/metal oxide very stable
    ii Molten electrolyte
    iii Electricity is expensive
4 a Haematite
  b High temperature
  c Coke, limestone, hot air
  d i $2C + O_2 \rightarrow 2CO_2$
    ii $Fe_2O_3 + 3CO \rightarrow 2Fe + 3CO_2$
    iii $Fe_2O_3 + 3C \rightarrow 2Fe + 3CO$
  e i $CaCO_3 \rightarrow CaO + CO_2$
    ii SiO_2
    iii $CaO + SiO_2 \rightarrow CaSiO_3$
    iv CaO is basic, SiO_2 is acidic
    v Foundations of buildings and roads
  f i Carbon, sulphur, phosphorus (or manganese, silicon)
    ii Oxygen is blown into molten iron and the impurities are converted into gaseous oxides (blown away) or liquid oxides (float as slag on the iron); sulphur removed using magnesium
  g Pig iron is brittle – main uses are ornamental; steel is hard and strong

62 Aluminium and titanium

1 a It forms a carbide: $TiO_2 + 3C \rightarrow TiC + 2CO$
  b $WO_3 + 3H_2 \rightarrow W + 3H_2O$
2 a Bauxite
  b Electricity too expensive to waste passing through impurities
  c Molten
  d To lower the melting point and improve conduction
  e Graphite
  f i $Al^{3+} + 3e^- \rightarrow Al$
    ii $2O^{2-} \rightarrow O_2 + 4e^-$
  g Oxygen oxidises the carbon anode so it has to be replaced periodically
  h Cheaper hydro-electric power
  i Low density – alloys in aerospace industry; resists corrosion – food packaging; good conductor of electricity – overhead power cables
  j Three from: save energy (high cost of production), maintain reserves of ores, maintain other reserves e.g. coal, water, less waste to dispose of, reduce air pollution
3 a Rutile
  b $TiO_2 + 2C + Cl_2 \rightarrow TiCl_4 + 2CO$ heat, dry
  c i Sodium, magnesium
    ii Argon atmosphere, heat
    iii $TiCl_4 + 4Na \rightarrow T + 4NaCl$
  d Chlorine and sodium both have to be produced first – by electrolysis, both stages require hot, dry conditions to prevent hydrolysis of $TiCl_4$, argon atmosphere to prevent oxidation
  e i Low density, strong
    ii Does not corrode
    iii Impurities make it brittle
  f Suitable ionic solvent to lower the melting point of TiO_2 and improve its conduction, could then extract larger quantities by electrolysis; titanium stronger and more resistant to corrosion than iron and steel
4 a Continuous process – does not have to be stopped to recharge it; batch process – has to be stopped to recharge it, so more expensive
  b Iron and aluminium – continuous; titanium – batch

63 Methods of extraction

1 a Very reactive metal – no suitable chemical reducing agent
  b To lower the melting point
  c Solid salts would not conduct electricity

Answers

    d i Cathode
        ii Anode
    e i $Na^+ + e^- \rightarrow Na$
        ii $2Cl^- \rightarrow Cl_2 + 2e^-$
    f They react together
    g In aqueous solution sodium would react to give sodium hydroxide
    h Large deposits of salt
2 a $2ZnS + 3O_2 \rightarrow 2ZnO + 2SO_2$
    b SO_2 is a pollutant
    c $ZnO + C \rightarrow Zn + CO$
    d CO is toxic
    e $ZnO + CO \rightarrow Zn + CO_2$
    f Distillation
    d SO_2 used to make sulphuric acid for which the demand was high

65 Formulae of organic compounds

1 A-2, B-5, C-3, D-6, E-1, F-4
2 C_2H_6O
3 a CH_3CH_2CHO, CH_3COCH_3 (as structural formulae). Propanal, propanone
    b $CH_2=CH-O-CH_3$ (as a structural formula)
    c Two structures containing a ring. Correct structural formulae for methyloxacyclopropane and oxacyclobutane. Correctly drawn cis and trans isomers of $CH_3CH=CHOH$.
    d $CH_2=CHCH_2OH$ prop-2-en-1-ol and $CH_2=C(OH)CH_3$ prop-2-en-2-ol

66 Structural isomerism

1 a g/100g (H) = 11.765, moles (C) = 7.353, mole ratio = 5:8, $(C_5H_8) \times 1 = 68$, molecular formula = C_5H_8
    b Two rings in one structural isomer
    c Correctly draw structure of methylcyclobutadiene
    d Correctly drawn cis and trans isomers of $CH_2=CH-CH=CH-CH_3$
2 a All are isomers
    b C_5H_{10}
    c Correctly drawn structural formulae of methylcyclobutane (this isomer does not react in the Baeyer test), $(CH_3)_2C=CHCH_3$ (no geometrical isomers), $CH_3CH=CHC_2H_5$ (cis and trans isomers).
3 a $C_nH_{2n+2}S$, where $n = 1, 2, 3, 4 ...$
    b Sensible scales used on y and x axes. Draws smooth curve. As chain length increases surface area of contact between molecules increases, so stronger Van der Waals' forces of attraction to overcome.
    c Value from graph
    d $CH_3CH_2CH_2CH_2SH$, $CH_3CH(SH)CH_2CH_3$ drawn as structural formulae
    e $CH_3-S-CH_2CH_2CH_3$, $(CH_3)_3C-SH$ drawn as structural formulae

67 Empirical, molecular and structural formulae

1 $C_4H_8F_2$, C_2H_4F
2 Correctly drawn structures for:
    a $CH_3CH_2CH_2NH_2$
    b $(CH_3)_3C-CH(C_2H_5)_2$
    c $(CH_3)_3COH$
    d $CH_3CH(OH)CH_2CH(Cl)COOH$
    e correct cyclic structure
    f $HOCH_2CH(OH)CH_2OH$
    g correct cyclic structure
    h $ClCH_2CH(CH_3)CH_3$
    i $CH_3CH_2CH(CH_3)CH=CH_2$
    j $CH_3COCH(CH_3)_2$
    k correct cyclic structure
    l $ICH_2(CH_2)_4CH_2I$
    m $CH_2=C(CH_3)CH(CH_3)_2$
    n $NC(CH_2)_3CN$
3 Correctly drawn structural formula for each correct IUPAC name.
    a 2-methylpentane
    b 2-chlorobutane
    c 2-methylbutane
    d pentan-2-one
4 a The first two structures
    b The first and third structures
5 Correctly drawn structures for each of the following structural isomers: 1-bromopentane, 2-bromopentane, 3-bromopentane, 1-bromo-2-methylbutane, 1-bromo-3-methylbutane, 2-bromo-2-methylbutane, 2-bromo-3-methylbutane, 1-bromo-2, 2-dimethylpropane

68 Naming more organic compounds

1 a 3-chlorobutan-1-ol
    b 3-ethyl-4-methylhex-2-ene
    c 3-bromopentanal
    d 2-methylbutane
    e 2-hydroxypentanenitrile
    f 2-iodopropane
    g 3-chloro-2-methylpentane
    h 2-bromo-2-chloroethanoic acid
    i 3,4-dimethylpentanenitrile
    j prop-2-enal

69 Fractional distillation and cracking

1 Petroleum vapour
2 Differing boiling points
3 Four from: petrol/naphtha/kerosene/diesel/fuel oil/lubricating oil, with a use for each

Answers

4 Decreases
5 When vapours condense latent heat of vaporization is liberated
6 Those in fraction C are more highly branched than those in fraction D/more branching = more spherical/less contact between molecules. Weaker intermolecular forces
7 800–1000 K for thermal or 750–850 K + catalyst for catalytic/C—C bonds break/free radicals or ionic mechanism/smaller, more branched cyclic products formed
8 C—C lower bond energy/weaker

70 Industrial applications
1 a 2,2-dimethylhexane
   b 2,3,3-trimethylpentane
   c tetramethylbutane
   d 4-methylheptane
2 c, b, a, d/more branched/more spherical/less contact between molecules/ weaker intermolecular forces/lower boiling point
3 Chain
4 C—C bonds break first
5 Cracking
6 CH_3—$(CH_2)_{14}$—CH_3 = any alkane—alkene mixture adding up to $C_{16}H_{34}$
7 A group of compounds with similar chemical properties/gradually changing physical properties/same general formula/produced by similar chemical means/alkenes
8 $C_{16}H_{34} \rightarrow C_8H_{16}$ or $18 + C_6H_{12}$ or $14 + C_2H_4$ or 6 – H atoms must add up to 34/not enough hydrogen to bond to newly formed C terminals atoms/cannot all be alkanes unless C_nH_{2n} molecules are cyclic

71 Combustion of fuels
1 $C_8H_{18} + 12.5O_2 \rightarrow 8CO_2 + 9H_2O$ (or multiple of this)
2 Not enough oxygen
3 $C_8H_{18} + 8.5O_2 \rightarrow 8CO + 9H_2O$
4 CO/it combines with haemoglobin causing suffocation
5 Any named oxide of nitrogen/acid rain/ damage to environment or buildings/smog-forming/toxic
6 Catalytic converter
7 $2CO + 2NO \rightarrow 2CO_2 + N_2$ or two equations showing production of CO_2 and N_2
8 Greenhouse gas/global warming
9 Poisons the catalyst in the converter
10 Conserve oil reserves/a non-renewable resource/only make essential trips/walk when possible/share travel/use public transport
11 H_2SO_4
12 $S + O_2 = SO_2$/acid rain/damage to environment or buildings

72 Petroleum
1 800–1000 K
2 Homolytic fission of C—C bonds
3 Free radicals/unpaired e⁻
4 C—C bonds weaker than C—H bonds
5 750–850 K + catalyst
6 Ions/carboniun ions
7 Catalytic/lower temperature/cheaper
8 High octane/additives for petrol
9 $Cl_2 \rightarrow 2Cl^\bullet$ initiation
   $Cl^\bullet + CH_4 \rightarrow {}^\bullet CH_3 + HCl$
   ${}^\bullet CH_3 + Cl_2 \rightarrow CH_3Cl + {}^\bullet Cl$ both chain propagation
   ${}^\bullet Cl + {}^\bullet Cl \rightarrow Cl_2$ or ${}^\bullet Cl + {}^\bullet CH_3 \rightarrow CH_3Cl$ or
   ${}^\bullet CH_3 + {}^\bullet CH_3 \rightarrow C_2H_6$ chain termination
10 CH_3Cl
11 Always get a mixture of products/low yield

73 Unsaturated hydrocarbons
1 b and e are alkenes as they have C_nH_{2n} formulae
   a, c and d are alkanes as they have C_nH_{2n+2} formulae
2 Propene and butene

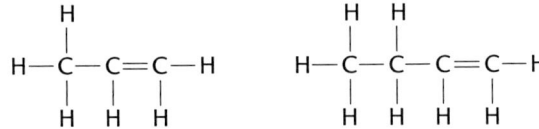

3 a C_2H_4
   b C_5H_{10}
4 a Double bond in different positions along the chain
   b Hex-3-ene and hex-2-ene
   c Both hex-1-ene
5 a Same molecular formular but different orientation of some atoms
   b Single bonds rotate, double bonds are rigid
   c cis-hex-3-ene and trans-hex-3-ene
6 a (cis) 3-methyl-pent-2-ene
   b $C_6H_{12} + 9O_2 \rightarrow 6CO_2 + 6H_2O$
7 b and d as both are alkenes (unsaturated)
8 Hexene: C_6H_{12}

74 Breaking the double bond
1 Addition reaction

Answers

2 **Double bond:** bond where two pairs of electrons are shared
Addition reaction: where one bond of a double breaks open to form new bonds
Electrophiles: species attracted to negative charge, i.e. with a + or δ+ charge/electron pair acceptor

3 a **Dipole:** particle with an uneven charge distribution: δ+ and δ- on each end of bond
Carbocation: a positive ion formed from and organic molecule, with the charge on a carbon atom
  b δ+ on HBr attracted to the double bond; carbocation and Br⁻ ion form; these then combine

4 a $CH_3CH=CH_2 + H_2SO_4 \rightarrow CH_3—CH(OSO_2OH)CH_3$
  b Sulphuric acid
  c CH_3CHCH_3
  d Propan–2–ol $CH_3CH(OH)CH_3$
  e Hydrolysis

5 a The electron cloud is spread evenly
  b A dipole is induced as the molecule approaches the high electron density (–ve) region of the C=C bond
  c Temporary dipole induced in chlorine molecule: this attacks the C=C bond forming a carbocation and a chloride ion: these combine to form dichloroethane (structural diagrams to show this should be included)

6 a 1, 2-dibromobutane from but-1-ene and bromine
  b Iodoethane from ethene and hydrogen iodide
  c Butan-1-ol/butan-2-ol from butene and sulphuric acid → hydrolysed to butanol

75 Major and minor products

1 a i Ethene
    ii Propene
    iii But-2-ene
    iv But-1-ene
  b ii and iv

2 a i Tertiary
    ii Secondary
    iii Primary
  b i, has three-electron releasing alkyl groups – stabilising effect on the cation

3 a A: 1-bromopentane, B: 2-bromopentane
  b B major, A minor: the secondary carbocation is the more stable

4 a React pent-1-ene with sulphuric acid and then hydrolyse the products
  b Appropriate structural diagrams – major and minor routes via hydrogensulphates to alcohols
  c Pentan-2-ol, pentan-1-ol

5 i 2-chlorobutane: but-2-ene as sole product (or but-1-ene as major product) reacted with HCl
  ii hexan-1-ol: hex-1-ene reacted with sulphuric acid then hydrolysed – minor product
  iii 2-ethyl-2-bromobutane: 2-ethyl-but-1-ene plus HBr as major product

76 Hydrogenation

1 a

$$\begin{array}{c} H \\ \diagdown \\ C=C \\ \diagup \\ H \end{array} \begin{array}{c} H \\ \diagup \\ \\ \diagdown \\ H \end{array} + H—H \rightarrow H—\underset{\underset{H}{|}}{\overset{\overset{H}{|}}{C}}—\underset{\underset{H}{|}}{\overset{\overset{H}{|}}{C}}—H$$

  b Hydrogenation (another addition reaction)
  c Nickel
  d It is the polar nature of HBr that causes the electrophilic effect, attacking the C=C bond.

2 a Saturated means all single bonds, unsaturated means one or more C=C bond (or triple bond).
  b Animal fats are saturated; vegetable oils are unsaturated.
  c The oils are hydrogenated using a finely divided nickel catalyst.

3 a An **addition polymer** is a long chain molecule made from many smaller alkene molecules (the **monomer**). The double bond opens up to form two single bonds which then link into a chain. The 'opened out' alkene is the **repeating unit**.

$$n \begin{array}{c} H \\ \diagdown \\ C=C \\ \diagup \\ H \end{array} \begin{array}{c} H \\ \diagup \\ \\ \diagdown \\ H \end{array} \rightarrow \left[\begin{array}{c} H \\ | \\ —C— \\ | \\ H \end{array} \begin{array}{c} H \\ | \\ C— \\ | \\ H \end{array} \right]_n$$

  b Suitable sequence
  c Much larger molecules in poly(ethene), hence larger van der Waals' forces between them

4 High density is stronger – the closer the molecules pack, the stronger the force of attraction between them

Answers

5 a

$$\left[\begin{array}{cc} H & H \\ | & | \\ -C-C- \\ | & | \\ H & CN \end{array} \right]$$

b Propenonitrile, $CH_2=CHCN$
c Poly(propenonitrile)

6 The free radical attacks and opens the double bond, leaving a lone electron on the other carbon atom. Therefore this itself becomes a free radical and can attack a second C=C bond, producing a longer chain free radical, and so on. Finally the propagating chain contacts another of the original free radicals, and the sequence is terminated.

77 Alcohols

1 Less labour intensive – easier to automate, therefore cheaper
2 See the text book, page 206 – figures to match headings given in question
3

$H-\underset{H}{\overset{H}{C}}-\underset{}{\overset{H}{C}}=\underset{}{\overset{H}{C}}-\underset{H}{\overset{H}{C}}-H + H-OH \rightarrow H-\underset{H}{\overset{H}{C}}-\underset{H}{\overset{H}{C}}-\underset{OH}{\overset{H}{C}}-\underset{H}{\overset{H}{C}}-H$

$H-\underset{H}{\overset{H}{C}}-\underset{H}{\overset{H}{C}}-\underset{}{\overset{H}{C}}=C\overset{H}{\underset{H}{\diagup}} + H-OH \rightarrow$

products giving $H-\underset{H}{\overset{H}{C}}-\underset{H}{\overset{H}{C}}-\underset{OH}{\overset{H}{C}}-\underset{H}{\overset{H}{C}}-H$

or $H-\underset{H}{\overset{H}{C}}-\underset{H}{\overset{H}{C}}-\underset{H}{\overset{H}{C}}-\underset{OH}{\overset{H}{C}}-H$

4 a Ether is usually ROR – in this case the alkyl part is joined to make a triangular bond ring
b This is a strained structure (60° bonds instead of the tetrahedral 109°)
5 Silver catalyst, 180°C

$\underset{H}{\overset{H}{\diagdown}}C=C\underset{H}{\overset{H}{\diagup}} + \frac{1}{2}O_2 \rightarrow H-\overset{O}{\overset{\diagup \diagdown}{C-C}}-H$ (with H H below)

$H-\overset{O}{\overset{\diagup \diagdown}{C-C}}-H + H-OH \rightarrow H-\underset{H}{\overset{OH\ OH}{\underset{|}{C}-\underset{|}{C}}}-H$

Hydrolysed

6 a $-OCH_2CH_2-$
b from epoxyethane
7 a Water molecules
b Addition followed by the elimination of a small molecule

78 Haloalkanes (1)

1 See text book, Unit 14

2 a 2-chloropropane
b 1,2-dibromobutane
c 1-iodo-3-methylbutane
d 1-bromo-3-chlorobutane
e 1,1,2-tribromobutane

3 a $CH_2=CH_2 + Br_2 \rightarrow CH_2BrCH_2Br$
b electrophilic addition

4 a Bromine, u.v. light
b $C_2H_6 + Br_2 \rightarrow C_2H_5Br$
c There may be further bromination

5 a 2-bromopropane $CH_3CHBrCH_3 + NaOH \rightarrow CH_3CH(OH)CH_3 + NaBr$
b 1-bromo-2-methylpropane $(CH_3)_2CHCH_2Br + NaOH \rightarrow (CH_3)_2CHCH_2OH + NaBr$
c 2-bromo-2-methylpropane $(CH_3)_2CBrCH_3 + NaOH \rightarrow (CH_3)_2C(OH)CH_3 + NaBr$
d bromoethane $CH_3CH_2Br + KCN \rightarrow CH_3CH_2CN + KBr$
e 2-bromopropane $CH_3CHBrCH_3 + KCN \rightarrow CH_3CH(CH_3)CN + KBr$
f bromoethane $CH_3CH_2Br + 2NH_3 \rightarrow CH_3CH_2NH_2 + NH_4Br$
Conditions: a, b, c: aqueous; d, e: aqueous ethanol; f: heat.

6 a butan-2-ol $CH_3CHBrCH_2CH_3 + NaOH \rightarrow CH_3CH(OH)CH_2CH_3 + NaBr$
b propene $CH_3CH_2CH_2I + KOH \rightarrow CH_3CH=CH_2 + KI + H_2O$
c 1-butylamine $CH_3CH_2CH_2CH_2I + 2NH_3 \rightarrow CH_3CH_2CH_2CH_2NH_2 + NH_4Br$
d butanenitrile $CH_3CH_2CH_2Br + KCN \rightarrow CH_3CH_2CH_2CN + KBr$
e 2,2-dimethylpropanenitrile $(CH_3)_3CI + KCN \rightarrow (CH_3)_3CCN + KI$
f butane-1,4-diol $Br(CH_2)_4Br + 2NaOH \rightarrow HO(CH_2)_4OH + 2NaBr$
g buta-1,4-diene $Br(CH_2)_4Br + 2KOH \rightarrow CH_2=CHCH=CH_2 + 2KBr + 2H_2O$
h cyclohexanol $(CH_2)_5CHBr + NaOH \rightarrow (CH_2)_5CHOH + NaBr$
i cyclohexene $(CH_2)_5CHBr + KOH \rightarrow (CH_2)_4CH=CH + KBr + H_2O$

79 Haloalkanes (2)

1 a $CH_3\overset{\delta+}{CH_2}-\overset{\delta-}{Br}$
b NH_3 OH^- $CH_3CH_2O^-$ CN^-

2 a $CH_3CH_2CH_2CH_2Br$ 1-bromobutane; $CH_3CH_2CHBrCH_3$ 2-bromobutane; $(CH_3)_2CHCH_2Br$ 1-bromo-2-methylpropane; $(CH_3)_3CBr$ 2-bromo-2-methylpropane
b $CH_3CH_2CH_2CH_2Br$ and $(CH_3)_2CHCH_2Br$
c $(CH_3)_2CHCH_2Br$ and $(CH_3)_3CBr$
d $CH_3CH_2CHBrCH_3$

© Farrow R, Gibbens D, Stirrup M and Vowles R, 2000. AS Chemistry for AQA – Resource Pack, Heinemann

Answers

e $(CH_3)_2CHCH_2Br + KCN \rightarrow (CH_3)_2CHCH_2CN + KBr$
$(CH_3)_2CHCH_2CN + HCl + 2H_2O \rightarrow (CH_3)_2CHCH_2COOH + NH_4Cl$
nucleophilic substitution; hydrolysis

3 iodoethane; C—I bond weaker, so it breaks more easily

4 **X** 1, 2-dibromoethane; **Y** butane-1, 4-dinitrile; aqueous ethanol
$BrCH_2CH_2Br + 2KCN \rightarrow NCCH_2CH_2CN + 2KBr$
nucleophilic substitution
$NCCH_2CH_2CN + 2HCl + 4H_2O \rightarrow HOOCCH_2CH_2COOH + 2NH_4Cl$ hydrolysis

5 a i 2-iodo-3-methylbutane
   ii nucleophile ($\rightarrow$ alcohol); base ($\rightarrow$ alkene)
   iii $(CH_3)_2CHCHICH_3 + 2NH_3 \rightarrow (CH_3)_2CHCH(NH_2)CH_3 + NH_4I$
  b i 1-iodo-2,2-dimethylpropane
   ii 2-iodo-2-methylbutane
   iii 3-iodopentane
   iv 2-iodopentane
   v 1-iodopentane; 1-iodo-3-methylbutane; 1-iodo-2-methylbutane

6 Your mechanisms must be accurately drawn; remember the essential points:
  a a curly arrow starts from a lone pair or a bond
  b a curly arrow arrives at **i** a δ^+ atom, to form a new bond, **ii** a positively charged atom, to give it a lone pair (O or N) or to form a new bond (C) or **iii** a single bond to form a double bond.

80 Haloalkanes (3)

1 **J** 1-bromo-1-methylcyclohexane; **K** 1-bromocyclohexanol

2 **L** 1-bromoethanol; **M** 2-hydroxypropanenitrile
$CH_3CH(OH)Br + KCN \rightarrow CH_3CH(OH)CN + KBr$
$CH_3CH(OH)CN + HCl + 2H_2O \rightarrow CH_3CH(OH)COOH + NH_4Cl$

3 $CH_3OCH_2CH_3$ $CH_3COOCH_2CH_3$

4 a **P** $CH_2BrCHBrCH_2Br$
   Q $CH_2(CN)CH(CN)CH_2CN$
   R $CH_2(COOH)CH(COOH)CH_2COOH$
   S $CH_2(OH)CH(OH)CH_2OH$
   T $CH_2=CBrCH_2Br$ **U** $CH_2=C=CHBr$
  b **P** 1,2,3-tribromopropane
   S propane-1,2,3-triol
   T 2,3-dibromopropene
   U bromopropadiene
  c nucleophilic substitution; hydrolysis; nucleophilic substition; elimination

5 a Cl_2 u.v. light; aqueous NaOH
  b Cl_2 u.v. light; KCN; HCl, heat
  c HBr; KCN; HCl, heat

6 **A** but-2-ene
  B 2-bromobutane
  C butan-2-ol
  D butanone
  E but-1-ene
  F 2-methylpropene
  G 2-bromo-2-methylpropane
  H 2-methylpropan-2-ol

81 Alcohols and dehydration

1 B = pentan-2-ol, F = dimethylpropanol, G = 3-methylbutan-1-ol
2 F distils off first/A left in flask
3 A, E, F and G
4 D
5 A and B/B and C/D and H/D and E/G and H
6 B and C
7 1-bromo-3-methylbutane/nucleophilic substitution/only one product
3-methylbut-1-ene/electrophilic addition/two products/H produced via secondary carbocation/G via primary/H major product/G minor
Therefore 1-bromo, 3-methylbutane better/higher yield of G
8 B/but-1-ene/cis/trans but-2-ene, plus structural diagrams

82 Elimination and other organic reactions

1 a T could be oxidized further so it must have an aldehyde group/so Q = CH_3—CH_2—CH=CH_2/R = CH_3—CH_2—CHBr—CH_2Br/S = CH_3—CH_2—CH(OH)—CH_2OH/T = CH_3—CH_2—CO—CHO/W = CH_3—CH_2—CO—COOH/X = CH_3—CH_2—CH_2—COOH/so P = primary alcohol/i.e. CH_3—CH_2—CH_2—CH_2OH

P $\rightarrow$ Q $CH_3CH_2CH_2CH_2OH \rightarrow CH_3$—$CH_2$—$CH_2$=$CH_2$ + H_2O
Q $\rightarrow$ R CH_3—CH_2—CH_2=CH_2 + $Br_2 \rightarrow CH_3$—CH_2—CHBr—CH_2Br
R $\rightarrow$ S CH_3—CH_2—CHBr—CH_2Br + 2OH$^- \rightarrow CH_3$—CH_2—CH(OH)—CH_2OH + 2Br$^-$
S $\rightarrow$ T CH_3—CH_2—CH(OH)—CH_2OH + 2[O] $\rightarrow CH_3$—CH_2—$\overset{O}{\underset{\|}{C}}$—CHO + 2$H_2O$

Answers

T → W $CH_3-CH_2-\overset{O}{\underset{\|}{C}}-CHO + [O] \rightarrow$
$CH_3-CH_2-\overset{O}{\underset{\|}{C}}-COOH$

P → X $CH_3-CH_2-CH_2-CH_2OH + 2[O]$
→ $CH_3-CH_2-CH_2-COOH$

b Conc. H_2SO_4/170°C
c Mechanism – protonation of OH/loss of H_2O/loss of H^+
d A and B + C and D   e D   f A and C

83 Ethanol production and reactions

1 $CH_2=CH_2 + H_2O \rightarrow CH_3CH_2OH$/strong acid/7000 kPa/ 300°C
$C_6H_{12}O_6 = 2CH_3CH_2OH + 2CO_2$/yeast/ 28°C/aqueous

2 Simpler plant = cheaper to build/maintain/ uses renewable resource/adds flavour

3 Fermentation/uses renewable resource/$CH_2=CH_2$ from oil may run out or become too expensive

4 Fermentation: 28°C optimum temperature for yeast/at higher temperature yeast is killed or enzymes denatured or less effective/Direct hydrolysis: exothermic/higher temperature = lower yield

5 $CH_3CH_2OH + [O] \rightarrow CH_3CHO + H_2O$
$CH_3CH_2OH + 2[O] \rightarrow CH_3COOH + H_2O$ or
$CH_3CHO + [O] \rightarrow CH_3COOH$

6 For CH_3CHO: add the alcohol drop by drop/ to the oxidizing agent/at a temperature above the boiling point of CH_3CHO but below that of CH_3CH_2OH/aldehyde distils off as it forms/ reducing the chances of further oxidation
For CH_3COOH: heat the oxidizing agent plus alcohol under reflux/ then fractionally distill

7 CH_3CHO: Fehling's solution or Tollen's/ result/no brick red or silver mirror with CH_3COOH
CH_3COOH: add $NaHCO_3$/gas given off/that forms a white precipitate with $Ca(OH)_2$ or limewater/no reaction with CH_3CHO

8 $CH_3-CH_2-CH_2OH = CH_3-CH=CH_2 + H_2O$/conc. H_2SO_4/170°C
$CH_3-CH=CH_2 + H_2O = CH_3-CH(OH)-CH_3$/dilute acid/140°C

9 $CH_3CH_2CH_2\overset{..}{O}H \underset{+H^+}{\overset{H^+}{\longrightarrow}} CH_3CH_2CH_2-\overset{+}{O}H_2$

$\underset{-H_2O}{\longrightarrow} CH_3CH-\overset{+}{C}H_2 \underset{-H^+}{\longrightarrow} CH_3CH=CH_2$
$\quad\quad\quad\quad |$
$\quad\quad\quad\quad H$

2nd mechanism
$CH_3CH=CH_2 \longrightarrow CH_3\overset{+}{C}H-CH_3 \longrightarrow$
$\quad\quad\quad \hookleftarrow H^+ \quad\quad \curvearrowleft :OH_2$
$CH_3-CHCH_3 \longrightarrow CH_3-CHCH_3 + H^+$
$\quad\quad |\quad\quad\quad\quad\quad\quad |$
$\quad HO-H\quad\quad\quad\quad OH$

10 When propene is hydrated propan-2-ol is the major product/propan-1-ol minor/only a small amount of propan-1-ol obtained

84 Reactions of epoxyalkanes

1 $CH_3-\overset{O}{\overset{/\ \backslash}{CH-CH_2}} + H_2O \rightarrow CH_3-CH(OH)-CH_2OH$

2 a $CH_3-CH(OH)-CH_2OH + 3[O] \rightarrow CH_3-CO-COOH + 2H_2O$
b ketopropanoic acid or pyruvic acid

3 a $CH_3-CO-COOH + H_2 \rightarrow CH_3-CH(OH)-COOH$/Ni catalyst
b 2-hydroxypropanoic acid or lactic acid

4 Conc. H_2SO_4

5 $CH_3-CH(OH)-COOH \rightarrow CH_2=CH-COOH + H_2O$/conc. H_2SO_4/170°C

6 Protonate OH/loss of H_2O/loss of H^+ from C_3

7 C=C: add Br_2(aq)/decolourised
COOH: add $NaHCO_3$/colourless gas/white precipitate with $Ca(OH)_2$(aq)

8 Flowchart with displayed structures to show:
$CH_3-\overset{O}{\overset{/\ \backslash}{CH-CH_2}} \rightarrow CH_3-CH(OH)CH_2OH \rightarrow$
$CH_3-CO-COOH \rightarrow CH_3-CH(OH)-COOH$
$\rightarrow CH_2=CH-COOH$

85 Classification and reactions

1 Primary/secondary/tertiary
2 Primary (aldehyde + carboxylic acid)
3 Tertiary
4 $CH_3-CH_2-CH_2-CH_2-CH_2OH/CH_3-CH_2-CH_2-CH(OH)-CH_3/CH_3-CH_2-CH(OH)-CH_2-CH_3$
5 $CH_3-CH_2-CH_2-CH_2-CH_2OH$
6 $CH_3-CH_2-CH_2-CH(OH)-CH_3$ and $CH_3-CH_2-CH(OH)-CH_2-CH_3$

7 Mechanism: protonate OH/loss of H_2O/loss of H^+/OH removed from one C atom/H removed from an adjacent C atom/
in $CH_3-CH_2-CH_2-CH_2-CH_2OH$ there is only one H that can be removed/
but in $CH_3-CH_2-CH_2-CH(OH)-CH_3$ and $CH_3-CH_2-CH(OH)-CH_2-CH_3$ there are H atoms on adjacent C atoms on both sides of the OH carbon atom

8 C 60.00/12 = 5.00/H 13.33/1 = 13.33/O 26.66/16 = 1.66/Empirical formula C_3H_8O/Molecular formula the same, as the alcohol is monohydric so it has only one O atom

9 $C_3H_8O + [O] = C_3H_6O + H_2O$ (propanone)

10 $CH_3-CH(OH)CH_3 \quad\quad CH_3-\overset{O}{\underset{\|}{C}}-CH_3$

Answers

Answers to student book 'test yourself'

Module 1

1. **a** Atomic radius increases down the group (1) since the outer electrons are in higher energy levels (1) that are further from the nucleus (1).
    Electronegativitity is a measure of the ability of an atom to attract the pair of electrons in a covalent bond (1): electronegativity decreases down the group (1): since the outer electrons (covalent pair) are further from the nucleus (1). *Total 6*
    b The Group II metal chlorides are all ionic except beryllium chloride which is covalent (1) since the Be^{2+} ion has a high charge/size ratio (1) and therefore is highly polarising (1): the Be^{2+} ion polarises chloride ions (1) to such an extent that beryllium chloride is covalent: beryllium chloride dissolves in non-polar organic solvents (1).
    Beryllium hydroxide is amphoteric (1) and will react as a base and as an acid whereas the other group II hydroxides are all basic. *Max. 5*

2. **a** Sodium, magnesium and aluminium all have metallic bonding (1): the strength of the metallic bonding increases as the size of the positive centres (ions) decrease (1) and the number of delocalised electrons increases (1). Hence melting point increases from sodium to magnesium to aluminium (1).
    Silicon has a giant (1) covalent (1) structure: a very large number of covalent bonds needed to be broken in order to break down the tetrahedral lattice and this results in a very high melting point (1). Phosphorus (P_4) (1), sulphur (S_8) (1) and chlorine (Cl_2) (1) are simple covalent (1) molecules: the melting points are low since the intermolecular forces are weak van der Waals' forces (1): the melting point increases as the mass of the molecules increases (1); hence the melting point increases from chlorine to phosphorus to sulphur (1). Argon exists as separate atoms (1) with weak van der Waals' forces (1). *Max. 12*
    b This is because the 4s sub-level fills (1) before the 3d sub-level (1) or Period 3 only involves the filling of the 3s and 3p levels (1) which together hold a maximum of 8 electrons. *Total 2*

3. **a** Energy required (1) to remove an electron (1) from a gaseous atom (1) (or in mole quantities). *Total 3*

    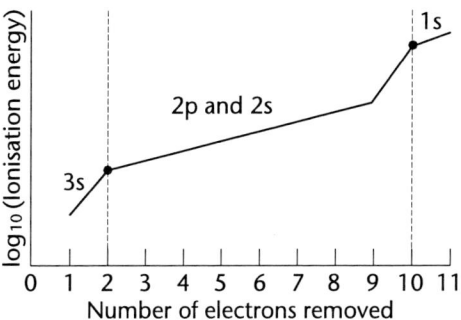

    b Successive ionisation energies always increase (1) in size since the remaining electrons (1) are more tightly held by the unchanged nuclear charge (1).
    Very large increases occur (1) when the electron removed is from a lower energy level (1). This occurs when the second (1) and tenth electrons (1) are removed from the sodium atom. *Max. 7*
    c $1s^2 2s^2 2p^6 3s^1$ (1) and $1s^2 2s^2 2p^6 3s^2$ (1)
    The outer electrons are in an s sub-level. (1) *Total 3*
    d i The sodium ion, Na^+ and Ne atom have the same electronic structure (1): the Na^+ has an extra proton (1) in its nucleus and the greater nuclear charge results in the second ionisation energy of sodium being greater (1) than the first ionisation energy of neon.
    ii The second ionisation energy of sodium involves removing an electron from the 2p sub-level (1): the second ionisation energy of magnesium involves removing an electron from the 3s sub-level which is further from the nucleus than the 2p (1) and thus more easily removed. Hence the second ionisation energy of magnesium is lower than the second ionisation energy of sodium. *Total 5*

4. Sodium chloride exists as a giant ionic lattice (1). A great deal of energy is required to overcome the ionic bonds (1) and break down

the lattice: hence sodium chloride has a very high melting point (1). When the ions are free to move then sodium chloride will conduct electricity (1): this occurs when it is melted or in solution (1).
Graphite is a giant covalent (1) structure (macromolecular). Each carbon atom forms three covalent bonds (1) to form layers of hexagonally (1) arranged carbon atoms. The fourth electron is delocalised (1) between the layers. A large number of covalent bonds need to be overcome to break down the lattice (1) and hence graphite has a very high melting point. The delocalised electrons make graphite a good conductor (1) of electricity.
Iodine exists as covalent (1) I_2 molecules (1). The forces of attraction between the molecules are weak van der Waals' forces (1) and hence it has a low melting point. There are no ions or delocalised electrons and therefore iodine is a non-conductor of electricity (1). *Max. 12*

5 From equation 1 mol of ammonium nitrite decomposes to form 1 mol of nitrogen and 2 mol of water in the gaseous state so 3 mol of gas are produced from 1 mol of ammonium nitrite (1). This means the total moles of gas produced from 0.01 mol of ammonium nitrite is 0.03 mol (1)
$$V = \frac{0.03 \times 8.31 \times (273 + 100)}{100 \times 1000} (1) = 9.3 \times 10^{-4} \, m^3$$
or 0.93 dm³ (1) *Total 4*

6 a

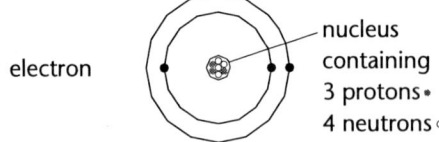
• electron
nucleus containing 3 protons
4 neutrons

Four marks for the diagram.

	Relative mass	Relative charge
Proton	1	+1
Neutron	1	0
Electron	1/1836	−1

One mark if both relative mass and charge are correct. *Total 7*

b Isotopes are atoms of the same element that have different number of neutrons in the nucleus (1). Chlorine has two isotopes ^{35}Cl and ^{37}Cl (1): both isotopes have 17 protons in the nucleus (1); ^{35}Cl has 18 neutrons and ^{37}Cl has 20 neutrons (1). The two isotopes have a natural abundance of $^{35}Cl:^{37}Cl$ of 3:1 (1). The mean relative atomic mass is 35.5. *Total 5*

c Ionisation: Electron gun (1) (or high energy electrons) knock an electron from an atom (1) forming a positive ion (1).
Acceleration: The positive ions are accelerated by an electric field (1): the positive ions pass through a slit (1) which produces a focussed beam (1) of positive ions.
Deflection: The positive ions are deflected by a magnetic field (1): the deflection depends upon the m/z (1) ratio of the ions (1): ions with a smaller m/z value are deflected more (1).
Detection: The positive ions are detected by an ion-current detector (1).
When a positive ion hits the detector an electron is released (1).
The size of the current is proportional to the number of positive ions hitting the detector (1).
The magnetic field is increased until ions with a particular m/z value are deflected and reach the detector (1). The magnetic field is then increased to allow ions with a greater m/z value to be deflected and reach the detector (1). *Max. 14*

d Moles of C 42.9/12 = 3.575
Moles of H 2.4/1 = 2.4
Moles of N 16.7/14 = 1.19
Moles of O 38/16 = 2.38
Two marks for the correct moles.
Simplest ratio (empirical formula) is found by dividing by 1.19.
This gives a ratio of C:H:N:O of 3:2:1:2 (1) and an empirical formula $C_3H_2NO_2$ (1).
Empirical formula mass = 84
From spectra M_r = 168 (1)
Hence molecular formula is $C_6H_4N_2O_4$ (1) *Total 6*

7 a CH_4 Carbon atom has four pairs of electrons (1): tetrahedral (1).
BCl_3 Boron atom has three pairs of electrons (1): trigonal planar (1).
NH_3 Nitrogen atom has three bond pairs and a lone pair (1): total of four pairs of electrons: shape is pyramidal (distorted tetrahedral) (1). *Total 6*

b NH_2^- Nitrogen atom has two bond pairs and two lone pairs (1): shape is distorted tetrahedral or V-shaped (1). Lone pairs result in a reduced bond angle of 105° (1).
IF_4^+ Iodine has four bond pairs and one lone pair (1): shape is based on a triangular bipyramid (1).

Answers

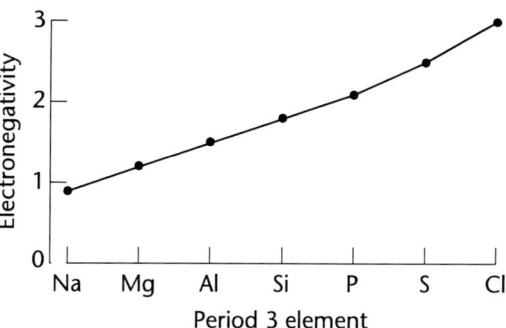

One mark for each correct shape.

c In NH_3 the nitrogen atom has three bond pairs and one lone pair (1): in H_2O, the oxygen atom has two bond pairs and two lone pairs (1). Both molecules have four pairs of electrons and have tetrahedral shapes (1). The extra lone pair in the water molecule result in greater distortion and a reduced bond angle (1). **Max. 15**

8 a From the equation the reacting ratio is 1 mole of Na_2CO_3 reacts with 2 moles of HCl (1).
Moles of acid = $(21.20 \times 0.113)/1000$ = 0.002396 (1)
Hence moles of Na_2CO_3 in $25\,cm^3$ is $0.002396/2 = 0.001198$ (1)
Hence moles of Na_2CO_3 in $250\,cm^3$ is $0.001198 \times 10 = 0.011998$ (1)
Hence mass of Na_2CO_3 in $250\,cm^3$ is $0.011998 \times 106 = 1.269\,g$ (1)

b Hence mass of H_2O in the sample is $2.995 - 1.269 = 1.726\,g$ (1)
Moles of $Na_2CO_3 = 1.269/106 = 0.011998$ (1)
Moles of $H_2O = 1.726/18 = 0.09588$ (1)
Simplest ratio of moles is 1:8 (divide both by 0.011998) (1)
Hence $Na_2CO_3.8H_2O$ (1) **Total 10**

9 a In water the oxygen atom forms two covalent bonds (1) with two separate hydrogen atoms. There are two lone pairs (1) of electrons on the oxygen atom (1). The shape of the water molecule is V-shaped (1) (or distorted tetrahedral). **Total 4**

b Below –10°C the water molecules vibrate (1) about a fixed position (1).
When the temperature reaches 0°C the molecules vibrate more strongly (1) and some of the forces between the molecules are loosened (1) allowing the molecules to move around (1). The ice has melted (1). When the temperature reaches 100°C the molecules have enough energy to overcome the forces of attraction between the molecules (1) and escape (1) from the liquid to form a gas. The water has boiled (1). In a gas the molecules move with rapid, random motion (1). **Max. 8**

10 a Atomic radius decreases (1) across the period since the outer electrons are all in the same principal energy level (1) but the nuclear charge steadily increases (1) and the electrons are drawn nearer to the nucleus (1). **Total 4**

b Electronegativity increases (1) as the atomic radius decreases (1) and the nuclear charge increases (1). **Total 3**

c i CsCl is ionic since the difference in electronegativity is large (1).
ii $BeCl_2$: the difference in electronegativity is still large (1); however the Be^{2+} ion is very polarising (1) and $BeCl_2$ is covalent.
iii NCl_3: the electronegativity values are the same and hence NCl_3 is covalent (1). **Total 5**

11 a $HCl + NaOH \rightarrow NaCl + H_2O$ (1)
b M_r of HCl is 36.5 (1)
Moles in 73 kg $73000/36.5 = 2000$ moles (1)
c This will react with 2000 moles of NaOH (1)
Concentration is 5 moles per dm^3
Hence volume needed is $2000/5 = 400\,dm^3$ (1).
d From the equation, 2 moles of HCl react with one mole of $Ca(OH)_2$ (1)
hence $2000/2 = 1000$ moles of $Ca(OH)_2$ needed (1).
M_r of $Ca(OH)_2$ is 74 (1)
Hence $(1000 \times 74)/1000 = 74\,kg$ of $Ca(OH)_2$ needed (1).
e Neutralisation with slaked lime is preferred (1). Sodium hydroxide is corrosive (1).
f One mole of $CaCO_3$ makes one mole of $Ca(OH)_2$ (1).
M_r of $CaCO_3$ is 100 (1)
100 kg of $CaCO_3$ will make 74 kg of $Ca(OH)_2$ (1)
Hence $(100 \times 5)/74 = 6.76\,kg$ of $CaCO_3$ will make 5 kg of $Ca(OH)_2$ (1). **Total 15**

Answers

12 a i Sodium and magnesium both have metallic bonding (1): the strength of the metallic bonding increases as the size of the positive centres (ions) decreases (1) and the number of delocalised electrons increases (1). The melting point depends on the strength of the metallic bond (1) so magnesium with the stronger metallic bonds has the higher melting point.

ii Silicon has a giant (1) covalent (1) structure: a very large number of covalent bonds needed to be broken in order to break down the tetrahedral lattice and this results in a very high melting point (1).

iii Sulphur (S_8) and chlorine (Cl_2) are both simple covalent (1) molecules: the melting points are low since the intermolecular forces are weak van der Waals' forces (1) but the melting point decreases as the mass of the molecules decreases (1); hence the melting point of chlorine is less than that of sulphur (1).
Max. 9

b i $\dfrac{1.78}{32}$ mol of sulphur combines with
$= 0.056$ mol
$\dfrac{0.89}{16} = 0.056$ mol of oxygen and $\dfrac{3.98}{35.5} = 0.112$ mol (1) of chlorine
S:O:Cl = 0.056 : 0.056 : 0.112
S:O:Cl = $\dfrac{0.056}{0.056} : \dfrac{0.056}{0.056} : \dfrac{0.112}{0.056}$ (1)
= 1 : 1 : 2 (1)
the empirical formula = $S_1O_1Cl_2 = SOCl_2$ (1)

ii The bonding will be covalent (1) because the three elements present are all non-metals (1).

iii $SOCl_2 + H_2O$ (1) $\rightarrow SO_2 + 2HCl$ (1)
Total 7

13 a The empirical formula of a compound gives the simplest ratio (1) of atoms of the elements present in the molecule (1) whereas the molecular formula gives the actual number (1) of atoms present. Total 3

b C : H = $\dfrac{85.72}{12} : \dfrac{14.28}{1} = 7.14 : 14.28$ (1)
C : H = $\dfrac{7.14}{7.14} : \dfrac{14.28}{7.14} = 1 : 2$ (1)
Therefore the empirical formula = CH_2 (1)

Using: $M_r = \dfrac{mRT}{pV}$ (1) where $m = 0.207$ g
$T = 273$ K (standard temperature)
$p = 1 \times 10^5$ Pa (standard pressure)
$V = 84 \times 10^{-6}$ m^3
$M_r = \dfrac{0.207 \times 8.31 \times 273}{1 \times 10^5 \times 84 \times 10^{-6}}$ (1) $= \dfrac{0.207 \times 8.31 \times 2}{1 \times 84 \times 10^{-1}}$
$= \dfrac{0.207 \times 8.31 \times 273 \times 10}{1 \times 84}$
= 55.91 which approximates to 56 (1)
the empirical formula mass = 12 + (2 × 1) = 14
$\dfrac{\text{Molecular mass}}{\text{Empirical formula mass}} = \dfrac{56}{14} = 4$ (1)
the molecular formula = $(CH_2)_4 = C_4H_8$ (1)
Max. 7

14 a i As the relative molecular mass increases the size of the molecule increases (1) so the points of contact between neighbouring molecules increases (1). This means the strength of the intermolecular forces between the molecules increases (1).

ii The boiling point of HF is higher than expected because the intermolecular forces between the molecules are hydrogen bonds (1). These are the strongest type of intermolecular forces (1). They arise due to the attraction of the $H^{\delta+}$ on one molecule with the $F^{\delta-}$ on an adjacent molecule (1) (or due to the donation of a lone pair from the F atom on one molecule to the $H^{\delta+}$ on an adjacent molecule). Total 6

b The boiling point of fluorine will be lower than that of HF (1) because fluorine is a non-polar molecule (1) and the only forces of attraction between the molecules are weak van der Waals' forces (1). Total 3

15 a 1 Ionisation (1)
Electron gun knocks an electron from an atom (1) forming a positive ion (1).

2 Acceleration (1)
The positive ions are accelerated by the electric field (1): the positive ions pass through the slit (1) (which produces a focussed beam (1) of positive ions).

3 Deflection (1)
The positive ions are deflected by the magnetic field (1): the deflection depends upon the m/z ratio of the ions (1).

Answers

4 Detection (1)
When a positive ion hits the detector an electron is released (1). The size of the current is proportional to the number of positive ions hitting the detector (1). (The magnetic field is increased until ions with a particular m/z value are deflected and reach the detector (1). The magnetic field is then increased to allow ions with a greater m/z value to be deflected and reach the detector (1).)

Max. 12

b i There are 3 peaks (1).

ii Similarities: They all have the same atomic number (1) and thus the same number of protons and electrons (1). They have the same number of outer electrons and thus the same chemical properties (1).

Differences: They have a different mass number (1) and thus a different number of neutrons (1): 12, 13 and 14 neutrons for the three isotopes (1). This means they have a different mass and thus vary in their physical properties (1), e.g density/rate of diffusion/melting point (1).

iii relative atomic mass = $\dfrac{\text{average mass of an atom}}{\text{mass of one atom of carbon-12}} \times 12$ (1)

iv A_r of magnesium =
$\dfrac{(24 \times 1) + (25 \times 0.127) + (26 \times 0.139)}{(1 + 0.127 + 0.139)}$ (1) (1)
= 24.32 (1)

Max. 12

Module 2

1 a The enthalpy change when one mole of a compound (1) is formed from its elements in their standard states (1) under standard conditions (298 K and 100 kPa) *Total 2*

b

$$6C(s) + 6H_2(g) \xrightarrow{\Delta H_f} C_6H_{12}(l)$$

with ΔH_1 down to $C_6H_6(l)$ and ΔH_2 up to $C_6H_{12}(l)$ (1)

$\Delta H_f = \Delta H_1 + \Delta H_2$
$\Delta H_f = +49 + (-208.1)$ (1)
$\Delta H_f = -159.10$ kJ mol^{-1} (1)

Total 3

c The enthalpy change when one mole of a substance (1) (element or compound) undergoes complete combustion (1) under standard conditions (298 K and 100 kPa) *Total 2*

d $C_2H_6O + 3O_2 \rightarrow 2CO_2 + 3H_2O$ (1)
Mass of ethanol that is burned (x) (1)
Volume (mass) of water (m) (1)
Temperature rise of the water (ΔT) (1)
Heat energy released $q = m \times 4.18 \times \Delta T$ (1)
(Specific heat capacity of water is 4.18 J g^{-1} K^{-1}) (1)
M_r of C_2H_6O is 46. Moles of C_2H_6O is $x/46$ (1)
Enthalpy of combustion is (heat energy released (q))/(moles of ethanol burned) (1)
Major source of error is heat loss (1) *Max. 7*

2 Very few collisions result in a reaction occurring (1) since only those collisions in which the particles have energy equal or greater than the activation energy (1) will result in a reaction occurring. The rate of a reaction can be increased by:
  i increasing the rate of collisions (1)
  ii increasing the percentage of the particles having energy equal to or greater than the activation energy (1).

a With the more concentrated acid there are more acid particles per unit volume (1): this results in an increase in the rate of collision (1); hence an increase in rate.

b The powdered calcium carbonate has a greater surface area (1) and hence there are more collisions per second (1) with the acid: hence an increase in the rate of reaction.

c When the solution is heated this results in more of the acid particles having energy greater than the activation energy (1): this results in an increase in the number of successful collisions (1). *Max. 9*

3 a One suitable method is to collect the oxygen gas formed in a graduated vessel such as a syringe (1). Suitable vessel (1). Suitable collection method (1).

b Graph: Moles (y-axis, from 2×10^{-4} to 20×10^{-4}) vs Time (sec) (x-axis, 0 to 200). Two curves shown, with sketch for part **(d)** being the steeper curve. * total no. of moles of oxygen produced = 0.00196

Axes (1), points correctly plotted (1), smooth curve (1).
Total number of moles of oxygen produced = 0.00196 (1)

c Moles of H_2O_2 = (0.20 × 20)/1000 = 0.004 moles (1)
Hence moles of oxygen = 0.004/2 = 0.002 moles (1)
Less oxygen is produced because some of the oxygen will dissolve in the water (1).

d See graph for part **b**.
Steeper curve (1) since increase in surface area gives faster rate (1). Same final amount of moles of gas produced (1) since same moles of hydrogen peroxide used (1). *Max. 13*

4 a

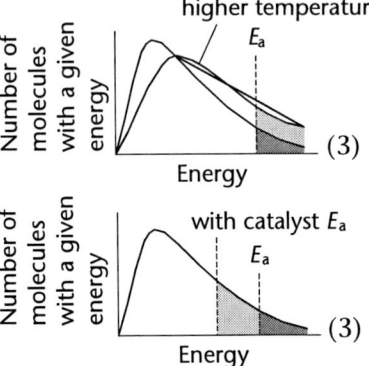

(3)

(3)

Answers

i When temperature is increased there are more particles with energy greater (1) than the activation energy (1) and hence the rate increases.

ii A catalyst provides an alternative reaction pathway (1) with a lower activation energy (1). Hence with a catalyst there are more particles with energy greater than the reduced activation energy (1).

iii A small increase in temperature produces a relatively large increase in the number of particles with higher energies (1) and hence a relatively large increase in the number of particles with energy greater than the activation energy.
A small increase in concentration only produces a small increase in the rate of collisions (1). *Max. 12*

b If some water is added to the reaction mixture then the equilibrium will shift to oppose the change (1). This will result in the equilibrium moving to the right (1) since this removes some of the added water (1). *Total 3*

c A catalyst has no effect on the position of the equilibrium (1) since it speeds up the forward and reverse reactions (1) equally (1). *Total 3*

5 a Homogeneous: all the reactants and products are in the same phase (state) (1).
Dynamic: the reactions continue to occur (1).
Equilibrium: both forward and reverse reactions occur at the same rate (1) and hence there is no overall change in the concentrations of the reactants (1). *Total 4*

b The forward reaction is exothermic (1) and hence the equilibrium shifts to the right-hand side at lower temperatures (1) so higher yields are obtained at lower temperatures (1). But at lower temperatures the rate of reaction is too slow (1). 450 °C is a compromise temperature (1). *Total 5*

c i The catalyst provides an alternative reaction route (1) which has a lower activation energy (1).

ii Finely divided gives the catalyst a greater surface area (1) and hence has a greater effect on the rate (1).

iii Catalyst speeds up the rate at which equilibrium is reached (1) but does not change the position of the equilibrium (1). Hence rate of production increases but the yield is the same (1). *Total 7*

6 a Oxidation is the loss (1) of electrons. In the equation: $Mg \rightarrow \frac{1}{2}O_2 + MgO$ magnesium is oxidised – oxygen is added. If the half equation for magnesium is considered: $Mg \rightarrow Mg^{2+} + 2e^-$ (1) it can be seen that the magnesium atom has lost electrons to form a magnesium ion. So oxidation is the addition of oxygen or the loss (1) of electrons. *Total 3*

b i As the atomic radius of the atoms increases (1) down Group VII the effect of the nuclear charge is reduced (1) due to more shielding so the halogen atom gains electrons less readily (1).

ii These are displacement reactions. Aqueous solutions of the halogens are added separately to aqueous solutions of the halide ions (1). The results are shown in the table below:

	Cl^- (aq)	Br^- (aq)	I^- (aq)
Cl_2(aq)	–	Yellow/orange solution (1)	Black ppt/ brown solution (1)
Br_2(aq)	No change (1)	–	Black ppt/ brown solution (1)
I_2(aq)	No change (1)	No change	–

Chlorine displaces bromine and iodine:
$Cl_2 + 2Br^- \rightarrow Br_2 + 2Cl^-$ (1)
$Cl_2 + 2I^- \rightarrow I_2 + 2Cl^-$ (1)
Bromine does not displace chlorine but does displace iodine:
$Br_2 + 2I^- \rightarrow I_2 + 2Br^-$ (1)
Iodine does not displace either chlorine or bromine. *Total 13*

c $Cl_2 + 2NaOH \rightarrow H_2O + NaCl + NaClO$
0   −1   +1
reduction (1)
oxidation (1)
A redox reaction involves oxidation and reduction. An increase in oxidation state is oxidation (1), a decrease in oxidation state is reduction (1). *Total 5*

7 a Silver nitrate solution added to potassium chloride solution produces a white precipitate (1) of silver chloride:
$Ag^+(aq) + Cl^-(aq) \rightarrow AgCl(s)$ (1)
Silver nitrate solution added to potassium iodide solution produces a yellow precipitate (1) of silver iodide:
$Ag^+(aq) + I^-(aq) \rightarrow AgI(s)$ (1) *Total 4*

© Farrow R, Gibbens D, Stirrup M and Vowles R, 2000. AS Chemistry for AQA – Resource Pack, Heinemann

Answers

b Concentrated sulphuric acid added to potassium chloride produces misty fumes (1) of hydrogen chloride:
$KCl(s) + H_2SO_4(aq) \rightarrow KHSO_4(aq) + HCl(g)$ (1)
Concentrated sulphuric acid added to potassium iodide produces purple fumes of iodine, a black precipitate of iodine, the smell of bad eggs due to hydrogen sulphide, a yellow precipitate due to sulphur: (2 marks for any 2 observations)
$KI(s) + H_2SO_4(aq) \rightarrow KHSO_4(aq) + HI(g)$ (1)
$8HI(aq) + H_2SO_4(aq) \rightarrow 4I_2(s) + H_2S(g) + 4H_2O(l)$ (1)
Total 6

c $Ag^+(aq) + Cl^-(aq) \rightarrow AgCl(g)$
    1.435 g
From the equation 1 mol of chloride ions produces 1 mol of silver chloride.
No. of moles of silver chloride formed = $\frac{1.435}{143.5}$
where 143.5 = M_r of AgCl (1).
No. of moles of AgCl = no. of moles (1) of Cl^- = 0.01 (1)
Mass of Cl^- in 0.01 mol = 0.01 × 35.5
    = 0.355 g (1).
% by mass of Cl^- in table salt = $\frac{0.355}{0.600} \times 100$
    = 59.2% (1)
Total 5

8 a Chlorides with concentrated sulphuric acid produce misty fumes (1) (of hydrogen chloride). Bromides with concentrated sulphuric acid produce brown fumes (1) (of bromine) and a choking odour (1) (due to sulphur dioxide). Iodides with concentrated sulphuric acid produce purple fumes (1) (of iodine) and a smell of bad eggs (1) (due to hydrogen sulphide) or a black and yellow precipitate mixture (due to iodine and sulphur). *Total 5*

b The reactions of bromides and iodides with concentrated sulphuric acid are redox reactions:
$Br^- + H_2SO_4 \rightarrow HSO_4^- + HBr$
    reduction (1)
$2HBr + H_2SO_4 \rightarrow Br_2 + SO_2 + 2H_2O$ (1)
 −1   +6   0   +4
    oxidation (1)

$I^- + H_2SO_4 \rightarrow HSO_4^- + HI$
    reduction (1)
$8HI + H_2SO_4 \rightarrow 4I_2 + H_2S + 4H_2O$ (1)
 −1   +6   0   −2
    oxidation (1)

In both cases the reactions occur in two steps. The first step displaces the hydrogen halide. This is not redox. The second step is redox, as illustrated by the changes in oxidation states shown above. *Total 6*

c Sodium astatide will react with concentrated sulphuric acid and initially hydrogen (1) astatide will be produced:
$NaAt + H_2SO_4 \rightarrow NaHSO_4 + HAt$
HAt will be the strongest reducing agent of all the hydrogen halides and will reduce the sulphur in concentrated sulphuric acid to its lowest oxidation of −2, i.e. the reaction is similar to that of HI:
$8HAt + H_2SO_4 \rightarrow 4At_2 + H_2S + 4H_2O$ (1)
    hydrogen sulphide (1)
Total 3

9 a i As the size of the halide ions increases (1) down Group VII the outer electrons are further away from the influence (1) of the nuclear charge. This means the ions are able to lose electrons more easily (1).

ii Hydrogen bromide reduces the sulphur in concentrated sulphuric acid from oxidation state +6 to oxidation state +4 (1) in sulphur dioxide (1). Hydrogen iodide reduces the sulphur in concentrated sulphuric acid further, resulting in sulphur (oxidation state = 0) (1) and hydrogen sulphide (oxidation state of sulphide is −2) (1). *Total 9*

b Electronegativity is a measure of the relative ability of an atom to attract (1) the pair of electrons in a covalent bond (1). The trend in electronegativity down Group VII is that it decreases (1). This is because down Group VII the halogen atoms increase in size (1) so the effect of the nuclear charge is reduced (1) by shielding. This means the ability of the atoms to attract electrons is reduced.
Total 5

10 All metals extractions are reduction reactions (1).
a Iron: raw materials are iron ore, e.g. haematite (1), coke (1) and limestone (1). Carbon monoxide is formed in the furnace. Carbon monoxide is the reducing agent (1).
$Fe_2O_3 + 3CO \rightarrow 2Fe + 3CO_2$ (1)
or $Fe_2O_3 + 3C \rightarrow 2Fe + 3CO$ *Max. 5*

b Aluminium: raw material is purified alumina from bauxite (1). Electrolysis of Al_2O_3 using carbon electrodes (1). Aluminium oxide is dissolved in molten cryolite (1) at 900 °C (1).

Answers

At the negative electrode: $Al^{3+} + 3e^- \rightarrow Al$ (1). This is reduction (1). *Max. 5*

  c Titanium is extracted using a more reactive metal (1) which acts as a reducing agent. Titanium chloride is reduced by sodium or magnesium (1) under a blanket of argon (1) at an initial temperature of 500 °C (1):
$TiCl_4 + 4Na \rightarrow Ti + 4NaCl$ (1)
or $TiCl_4 + 2Mg \rightarrow Ti + 2MgCl_2$ (1)   *Max. 5*

11 a Method of extraction of a metal is determined by: required purity of the metal (1), cost of the reductant used (1), cost of energy requirements (1).
Aluminium is extracted by electrolysis (1). Electrolysis of Al_2O_3 using carbon electrodes (1).
Aluminium oxide is dissolved in molten cryolite (1).
At the negative electrode:
$Al^{3+} + 3e^- \rightarrow Al$ (1)
At the positive electrode:
$2O^{2-} \rightarrow O_2 + 4e^-$ (1)
Titanium is extracted by reduction using a more reactive metal (1).
The titanium oxide is converted into titanium (IV) chloride by heating with carbon and chlorine at 900 °C (1).
$TiO_2 + 2C + 2Cl_2 \rightarrow TiCl_4 + 2CO$ (1)
Titanium chloride is reduced by sodium or magnesium under a blanket of argon (1) at an initial temperature of 500 °C.
$TiCl_4 + 4Na \rightarrow Ti + 4NaCl$ (1)
or $TiCl_4 + 2Mg \rightarrow Ti + 2MgCl_2$ (1)   *Max. 10*

  b When iron is extracted by carbon reduction it always contains some carbon (1). Small quantities of carbon improve the physical properties of iron (1). Steels contain small amounts of carbon (1). However, even small quantities of carbon will result in the formation of titanium carbide (1) which makes the metal brittle (1) and this means it has poor mechanical properties (1).
Extraction of iron can produce:
sulphur dioxide – acid rain (1), carbon dioxide – greenhouse gas (1), carbon monoxide – toxic (1).
Extraction of titanium can produce:
carbon monoxide – toxic (1), chlorine – toxic (1).   *Max. 10*

  c i The carbon monoxide reduces the iron oxide (1).
    ii M_r Fe_2O_3 = 160 (1).
Hence 160 kg of Fe_2O_3 makes 112 kg of iron (1).
Therefore 32 kg of Fe_2O_3 makes $(32 \times 112)/160$ (1) = 22.4 kg (1).   *Total 5*

12 a Excess potassium iodide is necessary to ensure that all the chlorine in bleach has reacted to release iodine: $Cl_2 + 2I^- \rightarrow I_2 + 2Cl^-$ (1). The end point is determined by adding starch solution (1). This produces a blue (1) coloration with the iodine which disappears at the end point when all the iodine has reacted (1) with the thiosulphate solution.   *Total 4*

  b From the equations 1 mol of ClO^- produces 1 mol of I_2 which reacts with 2 mol of $S_2O_3^{2-}$ (1).
No. of moles of $S_2O_3^{2-}$ = $\dfrac{15.4 \times 0.1}{1000}$
$= 1.54 \times 10^{-3}$ (1)
so the no. of moles of ClO^- used
$= \dfrac{1.54 \times 10^{-3}}{2}$ (1)
$= 7.7 \times 10^{-4}$ (1).
This is the no. of moles of ClO^- in 25 cm³ of diluted bleach. 10 cm³ of the original bleach was diluted to 250 cm³. So no. of moles in 10 cm³ of original bleach
$= 7.7 \times 10^{-4} \times 10$
$= 7.7 \times 10^{-3}$ (1).
Therefore no. of moles in 1.0 dm³ (1000 cm³)
$= 7.7 \times 10^{-3} \times 100$
$= 7.7 \times 10^{-1}$ (1)
Mass of sodium chlorate (I)
$= 7.7 \times 10^{-1} \times M_r$
$= 7.7 \times 10^{-1} \times 74.5$ (1)
$= 57.365$ g (1)
  Total 8

Module 3

1. **a** $CH_3CH_2CH_2CH_2OH$ butan-1-ol primary
   $CH_3CH_2CH(OH)CH_3$ butan-2-ol secondary
   $CH_3C(CH_3)(OH)CH_3$ 2-methylpropan-2-ol tertiary
   or $(CH_3)_3COH$
   $CH_3CH(CH_3)CH_2OH$ 2-methylpropan-1-ol primary
   or $(CH_3)_2CHCH_2OH$
   One mark for each. *Total 12*

   b butan-1-ol → $CH_3CH_2CH_2CHO$ (1) butanal (1)
   butan-2-ol → $CH_3CH_2COCH_3$ (1) butanone (1)
   2-methylpropan-2-ol → no change (1)
   2-methylpropan-1-ol → $CH_3CH(CH_3)CHO$ (1) 2-methylpropanal (1)
   These changes are oxidation reactions (acidified sodium dichromate (VI) solution is an oxidising agent). Primary alcohols are oxidised to aldehydes (1). Secondary alcohols are oxidised to ketones (1). Tertiary alcohols are not oxidised under these conditions (1). *Total 10*
   Note: Abbreviated structural formulae are used above but full displayed formulae can be used as an alternative.

2. **a** Stereosomerism occurs when two or more compounds have the same molecular and structural formulae (1) but their molecules differ in the orientation of some atoms or groups of atoms (1). But-2-ene shows a type of isomerism called geometrical isomerism:

   cis but-2-ene (1)   trans but-2-ene (1)

   This type of isomerism arises due to restricted rotation about the double bond (1). *Total 6*

   b Mechanism: electrophilic addition

   carbonium ion intermediate (carbocation)
   Hint: make sure the curly arrows are accurately positioned.

   2 bromobutane (1)

   Since but-2-ene is symmetrical (1) there is only one possible carbocation ion intermediate and thus only one product (1). *Total 8*

3. **a i** Free radical substitution (1) illustrated by methane reacting with chlorine:
   $Cl_2 \xrightarrow{uv} 2Cl^\bullet$ initiation step (1)
   $Cl^\bullet + CH_4 \rightarrow CH_3^\bullet + HCl$ } propagation
   $CH_3^\bullet + Cl_2 \rightarrow Cl^\bullet + CH_3Cl$ } steps
   $CH_3^\bullet + Cl^\bullet \rightarrow CH_3Cl$ termination step (1)
   chloromethane (major product)

   ii Nucleophilic substitution (1) illustrated by bromethane reacting with potassium cyanide:

   propanenitrile

   b The two alcohols formed are:
   $CH_3CH_2CH_2CH_2OH$ (1)
   butan-1-ol (1)   (minor product)
   $CH_3CH_2CH(OH)CH_3$ (1)
   butan-2-ol (1)   (major product)
   The resulting mixture contains a greater proportion of butan-2-ol (1). This is because butan-2-ol is formed via a secondary carbocation ion (1) intermediate whereas butan-1-ol is formed from a primary carbocation ion (1) intermediate. Secondary carbocation ions are more stable (1) than primary carbocation ions and are therefore more likely to be formed (1) resulting in the major product. *Total 9*

   c $CH_3CH_2CH_2Br + KOH \rightarrow CH_3CH=CH_2 + KBr + H_2O$
   (1 mark for equation, 1 mark for the structure of the organic product)
   The hydroxide ions act as bases (1). They remove a hydrogen atom (1) from the haloalkane to form water. If the solvent was water rather than ethanol the hydroxide ions would act as nucleophiles (1). They would donate a lone pair of electrons (1) to the carbon atom bonded to (1) the halogen. *Total 6*

4. **a** A nucleophilic substitution reaction of the haloalkanes involves the replacement (1) of the halogen atom by a nucleophile. A nucleophile is a lone pair donor (1). Haloalkanes undergo this type of reaction because the carbon—halogen bond is polar (1). The halogen is more electronegative (1) than carbon. This means the carbon atom

Answers

is electron deficient (1) and thus susceptible to attack by a nucleophile (1).

Total 6

[Reaction scheme: bromoethane + :NH₃ (nucleophile) → :Br⁻ + protonated ethylamine intermediate; :NH₃ (base) → ethylamine (CH₃CH₂NH₂) + NH₄⁺]   (6)

Excess ammonia is required to prevent further substitution (1) since ethylamine could also act as a nucleophile.

Total 7

c X = CH_3CN (1)
ethanenitrile (1)
(Nitriles are hydrolysed to carboxylic acids)

Total 2

5 a Ethanol is a renewable (1) energy resource if it is produced by fermentation (1).

$CH_2=CH_2 + H_2O \rightarrow CH_3CH_2OH$ (1)

or $CH_2=CH_2 + H_2O \longrightarrow CH_3CH_2OH$ (1)

The conditions necessary for this reaction are: a pressure of 7 MPa, a temperature of 300°C, a catalyst, e.g. concentrated phosphoric acid (on a silica support) (2) (for any 2 answers)

The advantages of this process compared to fermentation are:
i   The reaction is a continuous process
ii   It is fast
iii The product is pure
iv There is a high yield of product (2) (for any 2 answers)

The disadvantage of this process compared to fermentation is that ethene is obtained from crude oil which is a finite (non-renewable) resource (1).

c $CH_3CH_2OH + 3O_2 \rightarrow 2CO_2 + 3H_2O$ (1)

Bonds broken:
C—H × 5 = 413 × 5
C—C × 1 = 347
C—O × 1 = 358
O—H × 1 = 464
O=O × 3 = 498 × 3
4728 kJ mol⁻¹
(1)

This is an endothermic process.

Bonds made:
O=C=O × 2 = 805 × 4   (1)
H—O—H × 3 = 464 × 6
           = 6004 kJ mol⁻¹ (1)

This is an exothermic process.
$\Delta H = 4728 - 6004$ (1)
$= -1276$ kJ mol⁻¹ (1)

Total 7

d If petrol contains sulphur then sulphur dioxide (1) will be produced when the petrol burns in air. Sulphur dioxide causes acid rain (1).

Total 2

6 a B is an alcohol (1) (secondary aromatic alcohol)
C is a haloalkane (1) (secondary aromatic haloalkane)

Total 2

b Step 1: Reagents – $NaBH_4$ (or H_2) (1)
Conditions – in water (1)
(nickel catalyst)
Reaction type – reduction (1)
Step 3: Reagents – KCN (or NaCN) (1)
Conditions – ethanolic (1)
Reaction type – nucleophilic substitution (1)
Step 4: Reagents – H_2O/HCl (1)
Conditions – acidic (1)
Reaction type – hydrolysis (1)

Total 9

c When Tollen's reagent (1) (ammonical silver nitrate solution) is warmed with an aldehyde (1) a precipitate of silver (1) (a silver mirror) is formed. With a ketone there is no observable change (1).

or

When Fehling's solution (1) is boiled with an aldehyde (1) the blue solution turns red (1) (a red precipitate of copper (I) oxide is formed).
With a ketone there is no observable change (1).

Total 4

7 i 2-bromopropane and potassium hydroxide
If the potassium hydroxide is aqueous (1) then the type of reaction taking place is nucleophilic substitution (1). The product of this reaction is propan-2-ol (1):

Structure: $CH_3CH(OH)CH_3$

or $CH_3CH(OH)CH_3$ (1)

If the potassium hydroxide is alcoholic (1) (ethanolic) the type of reaction taking place is elimination (1). The product of this reaction is propene (1).

Answers

Structure:

$$H-\underset{\underset{H}{|}}{\overset{\overset{H}{|}}{C}}-\underset{\underset{H}{|}}{\overset{\overset{H}{|}}{C}}=C\overset{H}{\underset{H}{\diagdown}}$$

or $CH_3CH=CH_2$ (1)

ii Propan-1-ol and acidified potassium dichromate (VI).
If the reactants are gently warmed and the product immediately distilled (1) then propanal (1) is formed:

Structure:

$$H-\underset{\underset{H}{|}}{\overset{\overset{H}{|}}{C}}-\underset{\underset{H}{|}}{\overset{\overset{H}{|}}{C}}-C\overset{\diagup O}{\underset{H}{\diagdown}}$$

or CH_3CH_2CHO (1)
If the reactants are refluxed (1) then the product is propanoic acid (1):

Structure:

$$H-\underset{\underset{H}{|}}{\overset{\overset{H}{|}}{C}}-\underset{\underset{H}{|}}{\overset{\overset{H}{|}}{C}}-C\overset{\diagup O}{\underset{OH}{\diagdown}}$$

or CH_3CH_2COOH (1)
In both cases the type of reaction is oxidation (1). **Total 15**

8 a

$$H-\underset{\underset{H}{|}}{\overset{\overset{H}{|}}{C}}-\underset{\underset{H}{|}}{\overset{\overset{H}{|}}{C}}-C=C\overset{H}{\underset{H}{\diagdown}}\quad (1)$$

but-1-ene (1)

$$\underset{H_3C}{\overset{H}{\diagdown}}C=C\overset{H}{\underset{CH_3}{\diagup}}\quad (1)$$

cis but-2-ene (1)

$$\underset{H}{\overset{H_3C}{\diagdown}}C=C\overset{H}{\underset{CH_3}{\diagup}}\quad (1)$$

trans but-2-ene (1)

Total 6

b

$$H-\underset{\underset{H}{|}}{\overset{\overset{H}{|}}{C}}-\underset{\underset{H}{|}}{\overset{\overset{H}{|}}{C}}-\underset{\underset{OH}{|}}{\overset{\overset{H}{|}}{C}}-\underset{\underset{H}{|}}{\overset{\overset{H}{|}}{C}}-H$$

(1) $H-O-SO_3H$ (1)

↓ (1)

$$H-\underset{\underset{H}{|}}{\overset{\overset{H}{|}}{C}}-\underset{\underset{H}{|}}{\overset{\overset{H}{|}}{C}}-\underset{\underset{O^{\oplus}}{|}}{\overset{\overset{H}{|}}{C}}-\underset{\underset{H}{|}}{\overset{\overset{H}{|}}{C}}-H$$

$H\quad H$ (1)

(1 mark for structure + $:OSO_3H^{\ominus}$
1 mark for curly arrow)

$$H-\underset{\underset{H}{|}}{\overset{\overset{H}{|}}{C}}-\underset{\underset{H}{|}}{\overset{\overset{H}{|}}{C}}-\underset{\oplus}{C}-\underset{\underset{H}{|}}{\overset{\overset{H}{|}}{C}}-H + H_2O$$

(1) (1)
$:OSO_3H^{\ominus}$

↓

$$H-\underset{\underset{H}{|}}{\overset{\overset{H}{|}}{C}}-\underset{\underset{H}{|}}{\overset{\overset{H}{|}}{C}}-C=C\overset{H}{\underset{H}{\diagdown}} + H_2SO_4\text{ regenerates}$$

(1)

but-1-ene

Total 7

9 a Homolytic fission involved the breaking of a bond when the electrons from the bond are shared between the two atoms involved (1),
e.g. $Cl_2 \rightarrow 2Cl^{\bullet}$ (1) or $Cl^{\bullet\bullet}Cl \rightarrow 2Cl^{\bullet}$
A free radical is a reactive species which possesses an unpaired electron (1),
e.g. $Cl^{\bullet}$ (1) (a chlorine atom). **Total 4**

b Thermal cracking produces a high percentage of alkenes (1), e.g. ethene, as well as useful short-chain alkanes (1), e.g. octane. **Total 2**

c The direct hydration of ethene with steam

$$\underset{H}{\overset{H}{\diagdown}}C=C\overset{H}{\underset{H}{\diagup}} + H_2O \rightarrow H-\underset{\underset{H}{|}}{\overset{\overset{H}{|}}{C}}-\underset{\underset{H}{|}}{\overset{\overset{H}{|}}{C}}-OH\quad (1)$$

in the presence of a catalyst (phosphoric acid) produces ethanol (1).

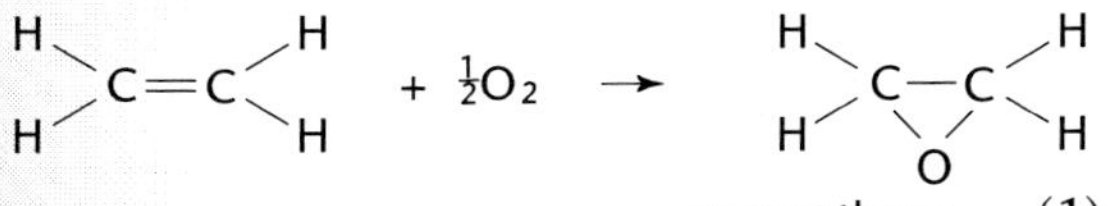

Answers

$H_2C=CH_2 + \frac{1}{2}O_2 \rightarrow$ epoxyethane (1)

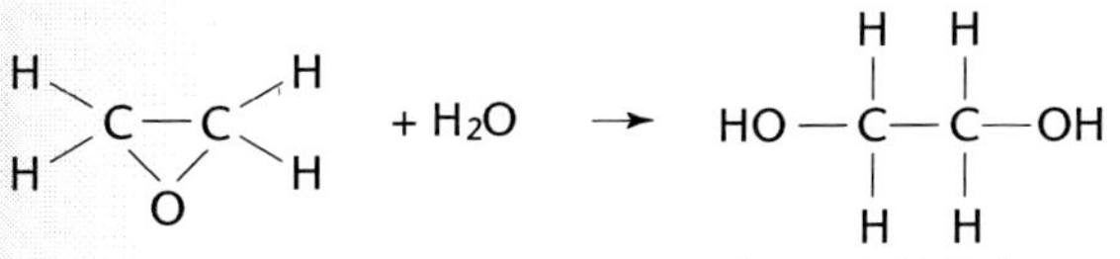

epoxyethane $+ H_2O \rightarrow$ HO–CH$_2$–CH$_2$–OH ethane-1,2-diol (antifreeze) (1)

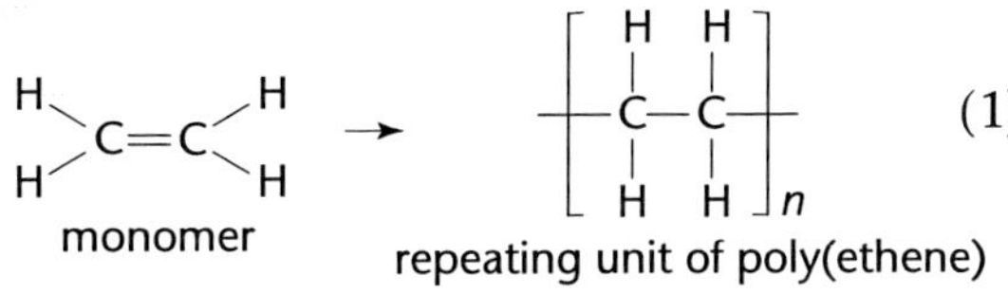

$H_2C=CH_2 \rightarrow$ [–CH$_2$–CH$_2$–]$_n$ repeating unit of poly(ethene) (1)

monomer

Hundreds of ethene molecules join together to form a long-chained polymer by addition polymerisation (1). **Total 6**

10 Propene → 2-bromopropane

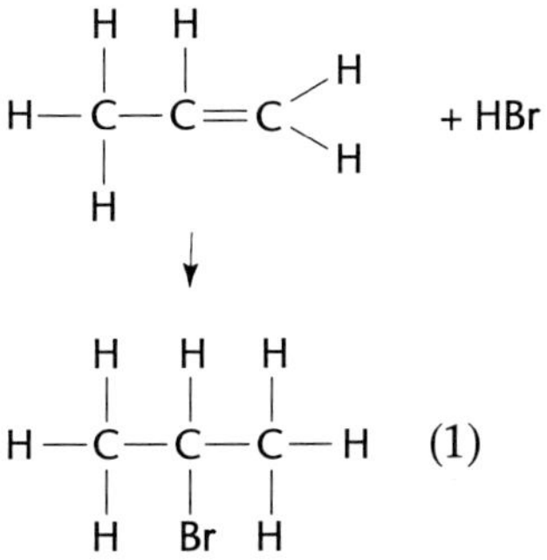

CH$_3$–CH=CH$_2$ + HBr → CH$_3$CHBrCH$_3$ (1)

or CH$_3$CH=CH$_2$ + HBr → CH$_3$CHBrCH$_3$

Reagent: hydrogen bromide
Type of reaction: electrophilic addition (1)
Mechanism:

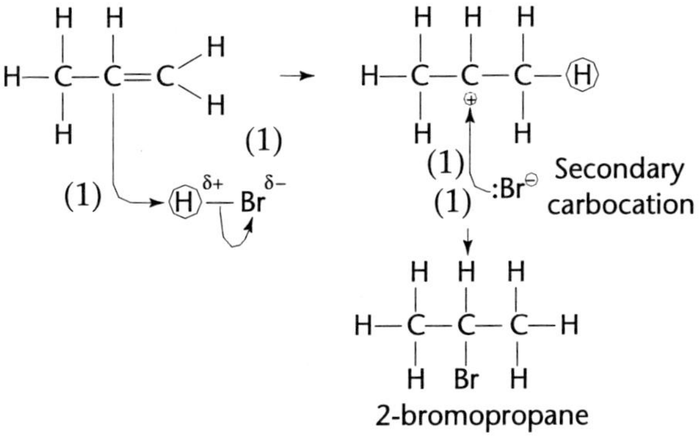

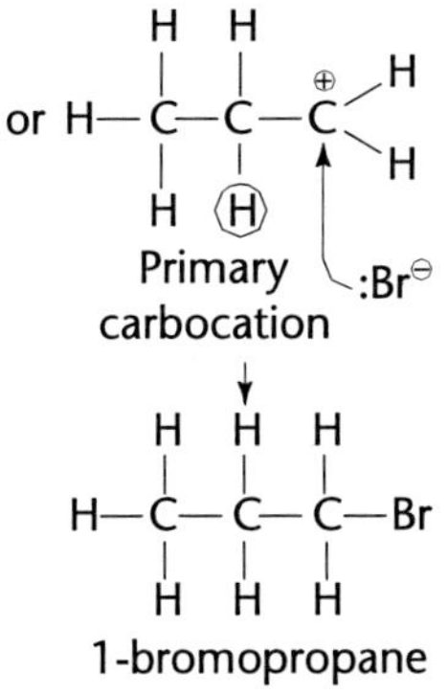

(mechanism showing secondary carbocation pathway leading to 2-bromopropane, and primary carbocation pathway leading to 1-bromopropane) (1)(1)(1)(1)

The major product, 2-bromopropane, is formed from a secondary carbocation which is more stable (1) than the other intermediate which is a primary carbocation ion (1). The latter will result in the minor product (1) which is 1-bromopropane.

2-bromopropane → propan-2-ol
Equation: CH$_3$CH(Br)CH$_3$ + OH$^-$ →
CH$_3$CH(OH)CH$_3$ + Br$^-$ (1)
Reagent: aqueous sodium hydroxide (1)
Type of reaction: nucleophilic substitution of hydrolysis (1)

Propan-2-ol → propanone (1)

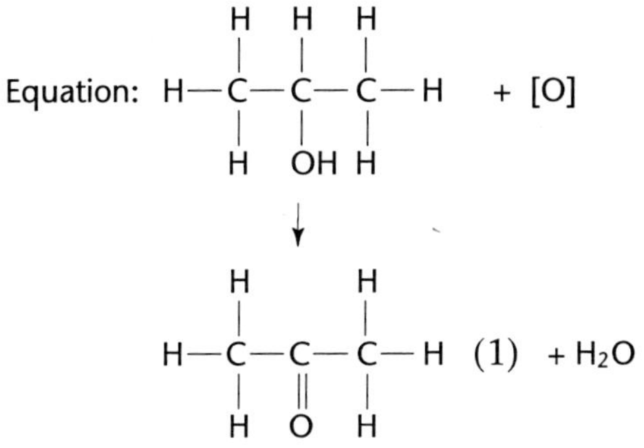

Equation: CH$_3$CH(OH)CH$_3$ + [O] → CH$_3$COCH$_3$ + H$_2$O (1)

or CH$_3$CH(OH)CH$_3$ + [O] → CH$_3$COCH$_3$ + H$_2$O

Reagent: acidified potassium dichromate(VI) (1)
Type of reaction: oxidation (1) **Total 15**

11 a 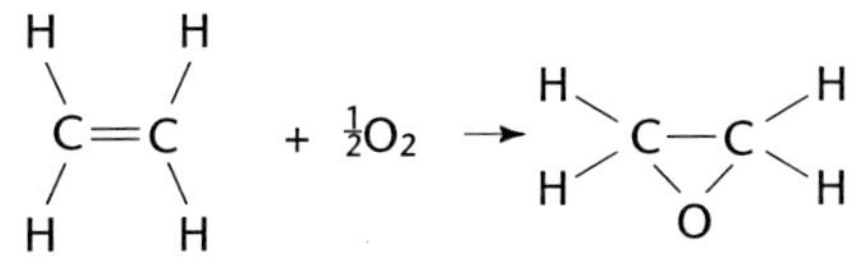
$H_2C=CH_2 + \frac{1}{2}O_2 \rightarrow$ epoxyethane (1)

Epoxyethane is manufactured by reacting ethene with oxygen in the presence of a silver catalyst (1).

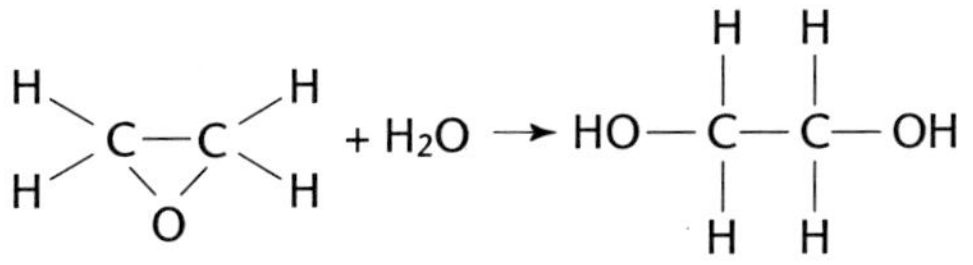

epoxyethane $+ H_2O \rightarrow$ HO–CH$_2$–CH$_2$–OH (1)

Epoxyethane is readily hydrolysed (1) by water to form ethane–1,2–diol.

b Antifreeze (1) or the production of polyesters (1).

c epoxyethane + CH$_3$CH$_2$OH → HO–CH$_2$–CH$_2$–OCH$_2$CH$_3$ (2)

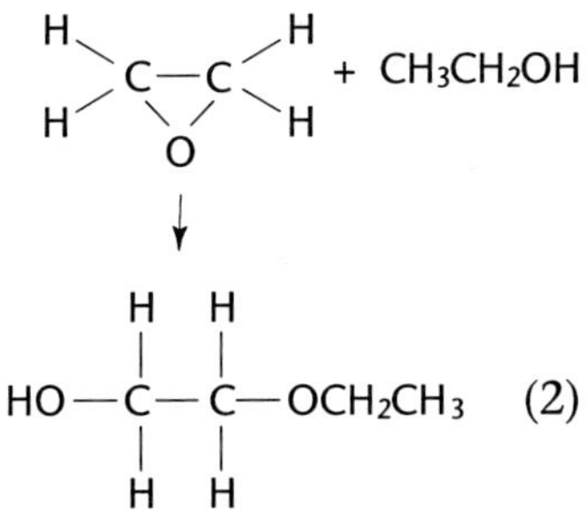

Answers

The product, which is an alkoxyalcohol, reacts with more epoxyethane to produce poly(alkoxyalcohols). These are used to make solvents, plasticisers and surfactants (2) (for any two uses).   Total 9

12 A carbocation is a species containing a positively charged carbon (1).

but-1-ene (1)

2-bromobutane

[Mechanism diagram showing but-1-ene reacting with H–Br (δ+/δ−), curly arrows, forming a carbocation intermediate with :Br⁻, then 2-bromobutane] (1) (1) (1) (1) (1)

butan-2-ol / but-1-ene

[Mechanism showing butan-2-ol with H–O–SO₃H (δ+/δ−), curly arrow, forming protonated intermediate + :ŌSO₃H, then carbocation + :ŌSO₃H, then but-1-ene + H₂SO₄ regenerates] (1 mark for structure, 1 mark for curly arrow)

but-1-ene

13 a Empirical formula of a compound shows the simplest ratio (1) of the atoms of the elements present. Molecular formula of a compound shows the actual number (1) of the different atoms present.   Total 2

b Molar ratio of C:H in hydrocarbon X
$$= \frac{85.7}{12} : \frac{14.3}{1}$$
$= 7.14 : 14.3$
$= 1 : 2$
Empirical formula = CH_2 (1) Empirical mass = 14
1 mole of any gas at stp occupies $22.4\,dm^3$
1 mole = the relative molecular mass in grams.(M_r)
If $85\,cm^3$ of gas has a mass of $0.21\,g$ then $22.4 \times 1000\,cm^3$ of gas has a mass of
$$\frac{0.21 \times 22.4 \times 10^3}{85} \;(1)$$
$M_r = 55.34$ (1)
$\frac{55.34}{14} = 3.95 \sim 4$
Molecular formula = $(CH_2)_4 = C_4H_8$ (1)
  Total 5

c

[Structure of but-1-ene] (1)
but-1-ene (1)

[Structure of but-2-ene] (1)
but-2-ene (1)

[Structure of 2-methylpropene] (1)
2-methylpropene (1)

Total 6

14 a An homologous series is a group of molecules which can be represented by the same general formula (1). The molecules have similar chemical properties and show a trend in physical properties (1).   Total 2

b i Alkanes are non-polar molecules (1) so the intermolecular forces between the molecules are van der Waals forces (1). As the molecules increase in size (1) so the points of contact with neighbouring molecules increases and the strength of the van der Waals forces increases (1).

© Farrow R, Gibbens D, Stirrup M and Vowles R, 2000. AS Chemistry for AQA – Resource Pack, Heinemann

Answers

ii The branch-chain isomer will have a lower boiling point (1). This is because since the molecule is more spherical (1) there will be less points of contact with neighbouring molecules so the van der Waals forces will be weaker (1).

iii Ethanol will have a higher boiling point than propane (1). Although the molecular masses (1) are similar, the intermolecular forces between the ethanol molecules are hydrogen bonds (1) which are stronger than van der Waals forces (1). **Max. 10**

15 a i The C—halogen bond in haloalkanes is polar (1) so the carbon atom carries a δ+ charge (1) making it susceptible to attack by nucleophiles.

ii The C—I bond in iodoethane is weaker (1) than the C—Cl bond in chloroethane so it is more easily broken (1). **Total 4**

b i Reagent = aqueous potassium hydroxide (1)
$CH_3CH_2CH_2CH_2Br + KOH \rightarrow CH_3CH_2CH_2CH_2OH + KBr$ (1)

ii Reagent = potassium cyanide (1)
$CH_3CH_2CH_2CH_2Br + KCN \rightarrow CH_3CH_2CH_2CH_2CN + KBr$ (1)

iii Reagent = ethanolic potassium hydroxide (1)
$CH_3CH_2CH_2CH_2Br + KOH \rightarrow CH_3CH_2CH=CH_2 + KBr + H_2O$ (1) **Total 6**

16 a Reagent = bromine water (1). With hexane the bromine solution remains orange (1). With hex-2-ene the bromine water is decolourised (goes colourless) (1). Formula of hex-2-ene is $CH_3CH_2CH_2CH=CHCH_3$ (1) **Total 4**

b Reagent = Fehling's solution (1). Heating is required (1). Butanal will turn the blue solution to a brick red (1). With butanone the solution will stay blue (1). Formula of butanal is $CH_3CH_2CH_2CHO$ (1) **Total 5**

c Reagent = sodium dichromate (VI) (1). The solution must be acidified and needs heating (1). Methylpropan-1-ol will turn the orange solution green (1). With methylpropan-2-ol (3° alcohol) the solution will remain orange (1). Formula of methylpropan-1-ol is $(CH_3)_2CHCH_2OH$ (1) **Total 5**

17 C_5H_{12} (pentane) shows chain isomerism (1). There are 3 isomers:

pentane (1)

2,2-dimethylpentane (1)

2-methylbutane (1)

C_3H_7Cl (chloropropane) shows positional isomerism (1). There are 2 isomers:

1-chloropropane (1)

2-chloropropane (1)

C_4H_8 (butene) shows positional (1) and geometrical (1) and chain isomerism (1):

but-1-ene (1)   but-2-ene (1)

2-methylpropene (1)

cis but-2-ene (1)   trans but-2-ene (1)

Answers

C_4H_8O shows functional group isomerism (1):

```
    H  H  H
    |  |  |    O
H—C—C—C—C⩘         (1)
    |  |  |    H
    H  H  H
```
butanal (1)

```
    H  H     H
    |  |     |
H—C—C—C—C—H   (1)
    |  |  ||  |
    H  H  O  H
```
butanone (1)

Max. 25

18 A = 2-bromobutane (1) $CH_3CH_2CHBrCH_3$ (1)
B = butan-2-ol (1) $CH_3CH_2CH(OH)CH_3$ (1)
The conversion of A to B is nucleophilic substitution (hydrolysis) (1).
C = butanone (1) $CH_3CH_2COCH_3$ (1).
The conversion of B to C is oxidation (1).
D = but-1-ene (1) $CH_3CH_2CH=CH_2$ (1) or but-2-ene (1) $CH_3CH=CHCH_3$ (1).
The conversion of A to D is elimination (1).
The reaction of D with concentrated sulphuric acid is electrophilic addition (1).
Max. 12

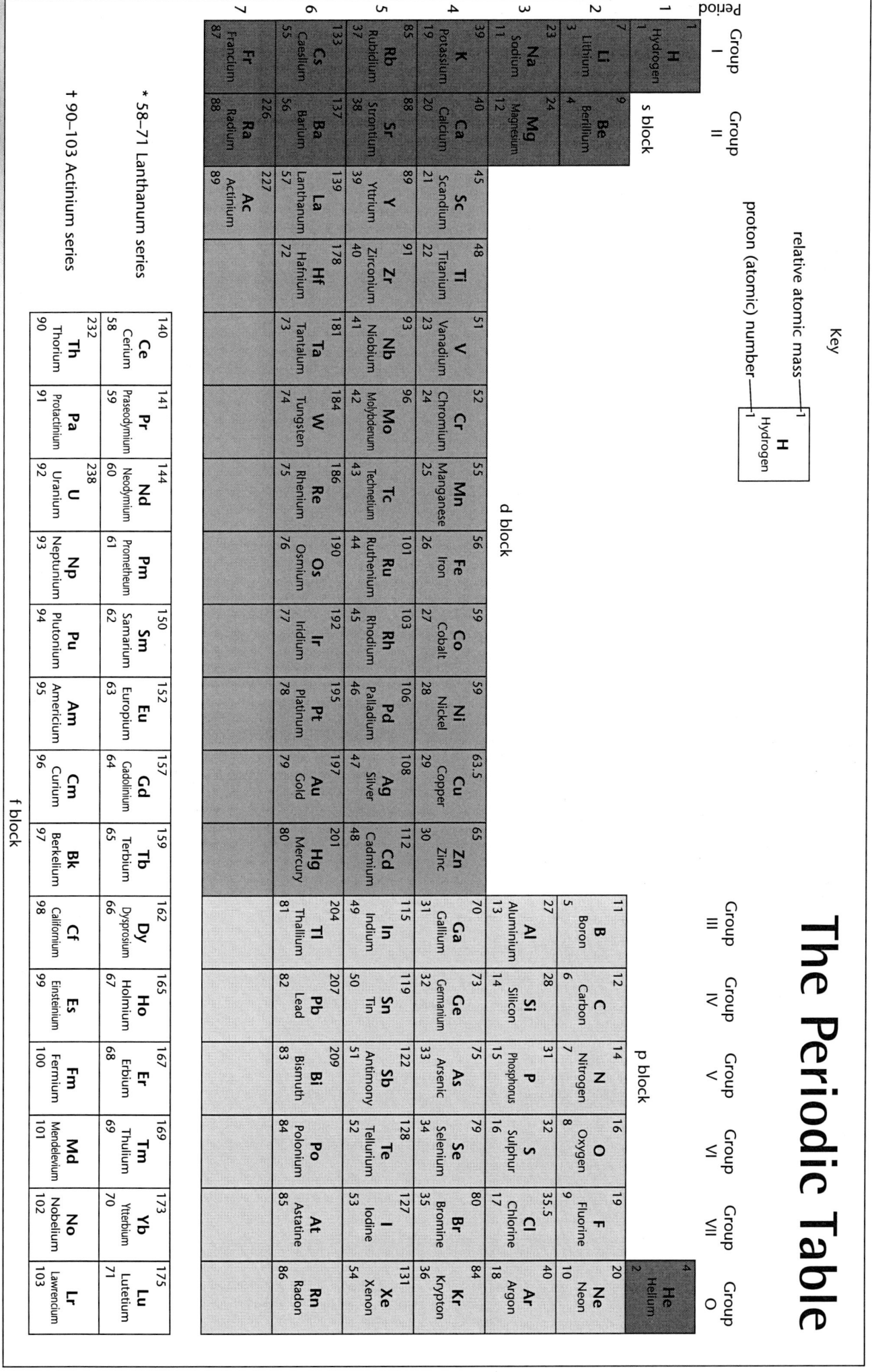